AQUACULTURE, LIMNOLOGY AND ICHTHYOLOGY

Manuals for Students, Researchers, Wildlife Managers and Environmentalists

AQUACULTURE, LIMNOLOGY AND ICHTHYOLOGY

Manuals for Students, Researchers, Wildlife Managers and Environmentalists

Outline drawn by
Prof. Hiralal Chaudhuri
Ex-Chief Technical Advisor, Aquaculture, FAO/UNDP

Edited by
A.B. Chaudhuri
Ex-Director, Forest Survey of India

2009
DAYA PUBLISHING HOUSE
DELHI-110 035

ISBN 81-7035-618-0
ISBN 978-81-7035-618-9

Published by : **Daya Publishing House**
1123/74, Deva Ram Park
Tri Nagar, Delhi - 110 035
Phone: 27383999
Fax: (011) 23260116
e-mail : dayabooks@vsnl.com
website : www.dayabooks.com

Showroom : 4760-61/23, Ansari Road, Darya Ganj,
New Delhi - 110 002
Phone: 23245578, 23244987

Laser Composed by : **Vaishnav Graphics & Systems**
New Delhi - 110 055

Printed at : **Chawla Offset Printers**
Delhi - 110 052

PRINTED IN INDIA

DEDICATION

This book is dedicated to the memory of Late Mukul Chaudhuri, wife of Professor Hiralal Chaudhuri, who had been his eternal source of inspiration and constant intellectual companion during his continued scientific success all through the service tenure.

ACKNOWLEDGEMENTS

Dr. S. B. Singh, Aquaculture Expert and Ex-Technical Advisor, FAO/UNDP deserves heartful thanks for his guidance in the preparation of the document. Thanks are also due to Shri B. K. B. Roy, IFS (Retd.), Ex-Principal Chief Conservator of Forests and Ex-Principal Secretary, Govt. of West Bengal, Department of Forests, for his precious cooperation in going through the manuscript and suggesting modifications here and there.

Professor Chaudhuri's Consent

Although I have always been averse to name, fame and public appreciation for whatever little I have achieved in my Scientific endeavour, I rather reluctantly permit Shri A.B. Chaudhuri of 131, N.S.C. Bose Road, Block-10, Flat 4, Kolkata-700 040 to edit my research documents, lecture notes, reports etc relating to aquaculture, limnology and ichthyology either in part or full or modify them for suitable publications.

All these papers were widely discussed, readout, quoted and presented in national and international Scientific Workshops, Symposia and Congregations. As such no accent is necessary for the present publication from any body or any organization.

H. Chaudhuri
Formerly Chief Technical Advisor, FAO/UNDP and
Regional Aquaculture Coordinator
South East Asian Fisheries Development Centre, Manila

Editor's request to readers

This book has been edited without any alive help from the Professor. Not only that, all documents are not available, hundreds of coloured transparencies could not be produced in the book. For brevity and to reduce expenses in production which has reduced the flavour of the manual. The Editor takes full responsibility for all lapses, mis or wrong presentation of facts if any.

Professor Chaudhuri's Vision

To meet the colossal demand of food in the present trend of population explosion, the development and Conservation of Inland and Marine water for food production is the only answer.

This indicates a diversion of attention from Agriculture to Aquaculture.

In fact the professor vowed in mid-fifties, or even earlier, much before the "Green Revolution" in Agriculture struck a success, to initiate a "Blue Revolution", or a Protein Revolution to create vast protein resources on a sustained basis to feed millions of people suffering from malnutrition and protein deficiency. His success in "Induced Breeding and introduction of various other techniques in aquaculture development testify to his practical and realistic approach to applied research and humaniterian services.

So the manuals outlined by him have immense practical value.

Editor

Foreword

The ingenuity and vision of M/s. Daya Publishing House, New Delhi have been reflected very well in their desire to record Professor Hiralal Chaudhuri's outstanding successes in the fields of Aquaculture, Limnology and Ichthyology researches, training and teaching in several publications. Professor Chaudhuri has been recognized as an internationally famous biologist of an high order. In fact he is the doyen among Aquaculture Scientists of the world. I congratulate the bold step taken by the publisher.

The selection of A.B. Chaudhuri, I.F.S. (Retd.) who has substantial contributions in the field of biology and forestry and to edit the Professor's published and unpublished technical works that brought about a "blue revolution" (Protein revolution) to feed the millions of poverty sticken people suffering from protein deficiency all over the world.

As for myself, I have closely studied the work of Professor Chaudhuri as I had the privilege to work with him for several decades at national and international arena – I was inspired by his devotion to research work as well as his kind approach to his juniors and subordinate. I consider those were my luckiest days to work with him on induced breeding of exotic major Chinese carps. His contributions to aquaculture for more than 50 years stand supreme as there was no end to demand for his technical advice from all over the world where he has left an indelible mark for monumental successes in all the projects he worked for. I am well acquainted with most of his technical works.

The magnanimity of the professor was of such a high magnitude that he considered all his successes have been due to combined effort of all the associates; and he has ever been averse to personal name and fame which he received felicitations and accolades of various top biologists from all over the world.

The editor has intended to prepare two treatise initially, one of the manuals, the professor prepared for his lecture in the University of Philippines and elsewhere; another on his technical publications on induced breeding, polyculture, nursery techniques, breeding etc. and also on unpublished ones.

The Professor had to be repeatedly persuaded for having his accent to edit on his work. It's a fact that the professor did not prepare the manuals for publication for easy communication of various facts of aquaculture, limnology and ichthyology but he prepared the manuals and used transparencies and colour slides, herbarium sheets and preserved specimens during lectures to students, researches and trainees.

The editor accepts his limitations and short comings in editing the manuals the outline of which was drawn by the professor. Most of his reprints and reports were donated to the Philippine University Library in 1993 and many, transparencies were damaged along with prepared manuals. So any gap found in the treatise must be due to obvious limitations of the editor, as he feels. However, I feel that the manuals edited by A.B. Chaudhuri based on outline draft of Professor Hiralal Chaudhuri provides enough facts expressed in lucid style for easy assimilation of the students, researches, trainees and for all other readers. These manuals are sure to build a solid foundation for study by the budding readers. The life-sketch and other achievements of the Professor as detailed in 'Preface' and 'Introduction' testify to the depth of knowledge and experience of the Professor and obviously the value of his creation of manuals though made in outline which is enormous.

I am sure the manuals will be very useful to the students, trainees, researchers, wildlife managers and environmentalists. In my view three manuals are quite similar to dictionary and grammer book, always needed in language classes.

I have personal pride, in all humility, while writing the foreword; the true feelings of a learner when citing the contribution of his guide.

S.B. Singh
F-2640, Palam Vihar,
Gurgaon-122017 Haryana, India

PREFACE

The buzzword *"Wetlands"* and *"Wetland Conservation"* received attention of both Central and State Governments in the decade 1990. This prompted the Governments in various States to form expert committees for documentation of available facts and figures related to Wetlands. This editor had the opportunity to be associated with a few committees in thier formative phases where some premier organizations viz, Bombay Natural History Society, Salim Ali Institute of Ornithological Studies, Bose Institute and several others were associated as members. The Institute which may claim as expert in Wetland research was ignored and no scientists either of State or Central Fisheries were involved as members.

I was astonished on such omission or say-ignorance about the monumental contribution in such wetland studies. Moreover during his various assignments in connection with EIA (Environmental Impact Assessment) relating to study of Aquaculture and Terrestrial Ecology he consulted Fishery Scientists' extensive published literatures in the fields of Physico-Chemical properties of water and soil, Phytoplankters, Zooplankters, Benthic biota, Productivity Energy dynamics and various other facts of wetland.

Moreover, their work concerning fish culture (breeding, hatchery, polyculture, genetics, biotechnology etc.) received international recognition and many fishery scientists (of Central inland fisheries) got international assignments of which Professor Chaudhuri stands supreme. The "Blue Revolution" in enhancing fish protein production obtained by success of induced breeding unearth, voluminous facts and figures about wetland flora, fish, water and soil. Several international organizations like FAO/ UNDP, IDRC, SEAFDEC and others highly felicitated the expertise and recognized their research findings.

Various countries of the world enhanced their fish production by applying the methdology and techniques introduced by Professor Chaudhuri. The Government of India was however, not conscious, of this revolutionary success in fish production until 2001 (i.e. after 43 years of initial success of hypophysation) when they declared 10th July, 2001 as "Fish Farmer's Day" all over the country which is to be observed every year.

M/s. Daya Publishing House, who have already published several books on Limnology and Fishery made a request to this editor to revise his book entitled "Wetland Ecology" published in 1998. This was done and the publisher published the book in 2006 entitled "Endangered Wetland".

Thereafter, M/s. Daya Publishing House made a request to this editor that they were interested to publish Professor Hiralal Chaudhuri's scientific works on Aquaculture, Limnology and Fisheries. It was accepted and initially two books were planned. But there were two major predicaments, firstly, the professor was averse to any publicity of his name and fame and he did not show any interests in to the proposal. He was of opinion that the successes were not exclusively due to him, but due to combined effort of all co-scientists who worked with him. There is no end to the magnanimity of the Professor whose life time activities were made with a missionary zeal for serving the poor, who did not get enough protein food and followed the age-old methods of Fishing.

Moreover it was found that most of his reprints, technical reports and unpublished papers were donated to the libraries of UPLB and ICLARM in 1993. As such, whatever few reprints and documents were available at his Kolkata residence were collected and analyzed to prepare this treatise for which the Professor gave permission rather reluctantly.

Of the two treatises planned, the first one (The present one) is a record of his success as a researcher more so as a teacher and a competent guide in the training of thousands of aquaculturists on the background of his monumental success in the field of aquaculture research.

The Professor had an inherent urge to become a teacher of 'Mathematics', but some socio-political situation in Assam forced him to take 'Biology' as Honours. He joined in the Biology Department of MC College, Sythet (now in Bangladesh) as a lecturer in 1943, but his tenure was short-lived. The partition of India threw out of the job and his dream of becoming a teacher was shattered.

The Professor got a berth in Central Inland Fisheries in 1948 and joined at Barrackpore as research assistant. But the Almighty God had another design. His sincere and arduous work took him to a position within nine years where he entered into a vast field of research and the success involved him in teaching and training arena at national and also at international levels.

His various national and international assignments, responsibilities as heads of many positions and projects made him a successful teacher, researcher, and a trainer in the vast fields of aquaculture, limology and ichthyology.

It was as such, considered a prime necessity to draw a short sketch of the professor's qualities to enable the readers to appreciate his work in preparing the three manuals.

During 1950 - 1953 his monumental work was in developing techniques to reduce heavy mortality of the young ones of carps during their early growing stage. The joint work was carried out by Mr. K. H. Alikunhi, H. Chaudhuri and V. Ramachandran. For this work Chandrakala Hora Gold Medal was jointly awarded to Mr. G. N. Mitra (Director-Organizer) and Mr. Alikunhi (Study Leader) for the 5-yr period 1951-1955.

In 1954 he was sent on deputation by GOI to USA under Point IV Program for 6 months initially, which was extended for another 6 months later on, obtained MS is Fisheries Management within the minimum period of 6 months.

He returned to CIFRS, (Central Inland Fisheries Research Station) Cuttack, succeeded in induced breeding of a carp minnow *Esomus danricus* by hormone injections.

He then continued induced breeding work and succeeded in breeding a riverine catfish *Pseudeutropius atherinoides* by pituitary gland injection.

In 1957 he was successful for the first time in the world when a member of the difficult-to-spawn Asiatic Carps, (*Cirrhinus reba*) was induced-bred on 10th july, 1957 by injections of carp hormones. This was immediately followed by the success in *Puntius sarana, Cirrhinus mrigala, Labeo rohita* and other Indian carps, and the technique was subsequently developed. The simple technique was followed for production of millions of carp seed. Training of high-ranked Fishery Officers from various states of India was initiated immediately after the success.

During 1957-1959 the training programs were continued and hundreds of State Fishery Officers, extension officers and even trainees were given the training in induced fish breeding.

Thereafter in 1958 he succeeded for the first time in the world in hybridizing carps by crossing the major and minor carps of the genera *Labeo, Cirrhinus* and *Catla* and obtained five inter-specific hybrids and eight-generic hybrids and a hybrid between rohu and common carp.

As Fisheries Extension Officer at Calcutta in 1959-1960 under his technical advice and guidance, millions of carp seeds were supplied to various States of India through the Fish Seed Syndicate with minimum mortality of seeds.

Back to CIFRS, Cuttack as Research Officer (Fish Breeding); in 1960-1961 he continued further research in induced breeding, conducting regular training courses and was invited to several states to establish fish breeding centers.

Outstanding Contributions in the Field of Aquaculture

Professor Chaudhuri did pioneering work in the field of Fish breeding and Fish culture in India, Burma, Fiji, Sudan, Nigeria and in a number of South-East Asian Countries. The more important among these are:

a) Induced Breeding of Indian major carps : Evolved and developed the technique of induced breeding of the commercially important major Indian carps by injection of fish pituitary hormones. This success has been considered as a scientific breakthrough in breeding the difficult-to-spawn cultivable Asiatic carps in ponds, a notable contribution in the field of Inland Fisheries in India. The technique developed, has been spread far and wide in the country and abroad and has contributed substantially to the growth of global aquaculture.

The know-how of this outstanding work has been imparted to Fishery Scientists and Fish farmers both in the country and from abroad through training courses accompanied by practical demonstrations, conducted under personal supervision and have helped in creating a trained cadre of fish breeders and aqua culturists. Most of the States of India and Burma are particular beneficiaries in having used this technique to attain self-sufficiency in the procurement of fish seed.

The techniques of large-scale production of the seed of the exotic Chinese carps were also demonstrated and the seed were distributed to various states in India and also to several countries abroad including Sudan, Fiji, Philippines, Nepal, Burma and Bangladesh. Success in induced breeding of the Asiatic carps has opened up a new field of research on Selective breeding and Hybridization.

b) Carp Hybridization : Attained success for the first time on record in hybridization of the cultivable carps, thus offering new possibilities towards evolving new strains and hybrids of superior qualities than the parents. Was able to produce hybrids exhibiting characters more desirable than the parent species. Fertile intergeneric hybrids were bred and F_2 and F_3 generations obtained.

c) Improved Nursery Management : Participated actively in the identification of the causes leading to heavy mortality of stocked carp fry in nursery ponds and evolved scientific management techniques in increasing average survival of fry from a meager 3% to over 60%. This contribution is of great significance in the production of fish for aquaculture.

d) Increased Production of Fish : The technique of polyculture of Asiatic carps has been developed at the Fish Culture Division at Cuttack under his guidance, supervision and participation through a decade of research resulting in high yields of 10 tons/ha/yr as compared to about 600 kg/ha/yr from fish ponds in India.

The success in the achievement of high yields through polyculture, prompted ICAR to initiate a Co-ordinated Research Project on Composite Fish Culture in various states under different agro-climatic conditions.

Under Operation Research/National Demostration Schemes, the technology proved highly successful in rural ponds, consequent to which a Project on Development of Rural Aquaculture in village ponds in West Bengal and Orissa, financed by IDRC, Canada was in operation under his guidance as the Project Co-ordinator. The technology thus developed may be considered a breakthrough in fish production in India and a few other countries where Asiatic Carps are cultured. As the FAO Expert in Burma, high production (over 10 tons/ha/yr) was demostrated by polyculture of carps.

e) Induced Breeding of Grey Mullet : Did pioneering work in successfully breeding *Mugil cephalus* by hypophysation in 1961 (First success in the world) at Chilka mouth.

f) Biological Control of Pistia stratiotes : Located a suitable Lepidopteran (insect) larva for the biological control of the noxious floating (Published in *Nature, 1975*).

g) Milkfish Breeding : Made substantial contributions towards successful induced breeding of milkfish *Chanos chanos* (First time in the world) at SEAFDEC in 1977, a major breakthrough in the field of world aquaculture. Also published for the first time the details of the embryonic development of milkfish and succeeded in distinguishing the sexes in milkfish spawners.

h) Integrated Farming of Crop-Fish-Livestock at Lao PDR : Demonstrated production over 4 tons/ha/yr production of fish through polyculture of Asiatic Carps integrate with duck/pig/poutry, without any additional inputs by way of artificial feeds and chemical fertilizers. The technique was followed throughout Laos which is one of the Least Developed Countries (LDC) of the world.

i) Research, Teaching and Training : The aforesaid achievements of the Professor on varied types of work very well justify his involvement in teaching, training and research which has been brought to light in this treatise.

More about his involvement has been presented in "Introduction" chapter.

A. B. Chaudhuri
131 N. S. C. Bose Road,
Block 10, Flat 4,
Kolkata-700 040.

Contents

BOOK III [illegible] ICHTHYOLOGY

Some Common Greek & Latin Prefixes & Common Words

Greek

A, An-Without

Amphi-Both

Ana-Up , Upon, Backward

Apo-Away from, Apart

Anti-Opposed to, Against

Di-Double, Twin

Dia-Throug, Across

Diplo-Double

Ec, Ek-Out of, Off

Ecto-Outside

En, Em-In, Into

Endo-Inside

Ento-Within

Epi-Upon

Haplo-Single, Simple

Hemi-Half

Hetero-Other

Homo-Same

Hyper-Above, Beyond, Greater than

Hypo-Under, Below

Gyno, Gyne-Woman, Female

Andro-Male

Trophein-To Nourish

Topos-Place

Allos-Other

Dys-Badly, Bad

Oikos (Eco)-House

Malakos-Soft

Miktos-Mixed, Mixing

Notos-Back

Rhis, Rhino-Nose

Rhynchos-Snout

Xeros-Dry

Oura, Uro-Tail

Iso-Equal, Similar

Lepto-Fine, Blender

Macro-Large, Big, Long

Mega, Megalo-Large

Mero-Part, Portion

Meso-Middle

Meta-After, Behind

Micro-Small

Mono-Single, Alone

Oligo-Small, Few

Ortho-Straight

Para-By the side of

Peri-Around, About

Poly-Many

Pro-Before

Proto-First

Sym, Syn-With, Together

Teleo-Complete

Tele, Telo-Distant, Far

Zygo-Unite, Yoke

Ptera-Wind

Eurys-Broad
Stenos-Narrow
Photos-Light
Helios-Sun
Thermos-Heat
Pelagos-High sea
Phyton-Plant
Podos, Pes-Foot
Rhodon-Red
Erythros-Red
Leuco-White
Melano-Black
Cyano-Blue Green
Aniso-Unequal
Archaeo, Archi-Ancient
Artho-Joint
Auto-Self
Brachio-Arm
Cata-Down
Cerato, Keras-Horn
Chele-Claw
Chiro-Hand
Chloro-Green
Cephalo-Head
Coel-Hollow
Dactylo-Finger
Derm, Dermatos-Skin
Echino-Spine
Eos-Dawn
Eu-Good, Well, True
Glosso-Tongue
Gnatho-Jaw
Gymno-Naked
Halo-Salt
Holos-Entire
Kryptos-Hidden
Ophio-Snake
Osteo-Bone
Stomos-Mouth
Proctos-Anus
Chilo-Lip
Lutea-Yellow
Phaga-To eat
Karyon-Nucleus
Physo-Bladder
Odonto-Tooth
Meion-Smaller, Less

Latin

a, ab-off, from, away
ad-to, at, towards, into
ant, ante-before
bi, bis-twice
co, com, con-with, together
de-from, away out, down
demi-half
e, ex-out of, without
extero-outer
extra-on the outside
infra-below, beneath
inter-Between, among
intra-within
ob-on the way of, against
per-through, by means of
post-after
pre-before
sub-under
super, supra-above
uni-one
littus, litus-shore
latero-side
lux, lucis-light
corpus-body
cerebro-brain
concho-shell
macula-spot
manus-hand
os-mouth-thread
pleion-more
pleistos-most
lepis, lepidos-scale
lignus-wood

Some Common Latin Words Used in Biological Science

a, ab.. off, from, away
ac... to, forward
acet.. vinegar
acetabul .. vinspar cup
ad .. tom at, toward, into
agni, agnus .. lamb
ala .. wing
alb, albine, . white
alis .. pertaining to
allo .. other, different
alti .. tail
ambi .. around
angui .. snake
ante .. before
arbor .. tree
arctic .. northern
areni .. sand
articulus .. joint
atri, atrius .. vestibule
aura .. gold
auris .. ear
austral .. southern
bi, bis .. two, twice
boreal .. north
brevi .. short
bucca .. cheek
caerule .. blue
cald, calid .. hot, warm
calix .. cup
cand, candid .. white brilliant
capill, capilus .. hair
capit, caput .. head
carni .. flesh
cauda .. tail
cauli .. a stem, stalk
ceps .. head
cerebro .. brain
cervi - x .. neck
catr, cetra .. shield
chalyb .. steel
cilium .. eyelid, fine hair
circum .. around

cirrus .. curl of hair
coccin .. scarlet
coli .. swell
com, con .. with, together
centra .. against
corn, cornus .. horn
corpus .. body
manus .. hand
maxilla .. jaw
medi .. middle
melano .. black
molli .. soft
multi .. many
nasi, nass .. nose
natau .. swim, flow
nudi .. naked
ob .. against, in the way of
cleva .. club
clavicle .. a key
crassi .. thick
crista .. crest
cucullus .. hood
de .. from, away, out, down
digitus .. finger
dis .. apart
dorsa, dorsum .. back
e, ex .. cut, without, from
ef .. out, from away
exter - o .. outer
extra .. on the outside
falc, falx .. sickle

fascia-tus .. banded, bundle
femus .. thigh
fer .. to carry, bear
ferrugin .. rust colored
filli, filum .. thread
flav .. yellow
fluvia .. river
foli .. leaf
font, fontan .. fountain
fusca .. brown
galb .. yellow
gallban .. greenish yellow
gero, ger .. to bear
gladius .. sword
glans .. gland
glaucc .. gray
gracil .. glender
grandi .. large
gravi .. heavy
hasta .. spear
homo .. man
llio .. intestine infra .. below, beneath
integri .. whole
inter .. between, among
intra .. within
labium, labrus .. lip
latero .. side
lati, laty .. wide, broad
lingua .. tongue
littor .. seashore, shore

longi .. long
lucid .. shining
luni .. moom
lute .. yellow
macula .. spot
stole .. a long robe
sub .. under
sulca .. groove
super, supra .. above
tarai .. ankle
tergum .. back
terti .. third
tetra .. four
tiana .. across
turbell .. a little crowd
ult .. beyond
oculus .. eye
ora, os .. smooth
os, ossi .. bone
ovus .. egg
pallid .. pale
palvi .. smell
pector .. breast
pedi, pes .. foot
per .. through, by mean of
pilus .. hair
pinna, pluma .. feather
pisei .. fish
post .. after
pre .. before
primi .. first
proxim .. near
pulmo - no .. lung
purpur .. purple
racti .. straight
remus .. oar
reni .. kidney
reti .. network
retro .. backwards
rostro, rostrum .. ancut, break
rubicund .. very red
rubra .. red
rugos .. wrinkle
sagitta .. arow
sanguini .. blood
saxa . rock
saxatil .. leaving among rocks
soutum .. shield
semi .. one-half
serra - t .. saw
sets, seti .. bristle
sol .. sun
soma, somato .. body
spini .. spine
squams .. scale
sterni .. breast unguis .. claw
uni .. one
ventro .. belly virid .. green
vora .. to eat
vittat .. striped
vulgur .. common

a, ab .. not, without
acantho .. spine, therm
aci, acu .. a point, needle
acro .. topmost, tip
acti, actine .. ray
adelo .. unknown, secret
adeno .. gland
adr, adro .. thick, stout
alg, algis .. poan
alveo .. hollow
amby .. blunt
ameb, ameba .. change
ana .. up, again, backward
andro .. man
angusti .. narrow
antho .. flower
anti .. opposed to against
aphan .. hidden, invisible
aphel .. smooth
apo .. away from, apart
arbor .. tree
arche .. ancient
archi .. first
argo .. shining, bright
argyr .. silver
arthro .. joint
asc, asco .. bag, bladder
aspi .. shield
astro .. star
auto .. self
auxo .. grow, enlarge

asc, azule .. dry, parched
badis, basac .. walk
bates .. one who haunts/walks
batho, bathy .. dep, high
belon .. needle, dirt
belu .. beast, moneter
benacus .. a deep lake
botry .. a bunch of grapes
brachium .. arm
brachy .. short
brady .. slow
branchium .. gill
broncho .. windpipe
byas .. fine thread
byth .. depth of sea
cal, calo .. beautiful
calypte .. cover, voil
campus .. sea monster
carchar .. jagged
dromo .. to run
dys .. bad, badly
ec, ek .. out of, off
echibno .. spine, hedgehog
ecto .. outside, outer
eido .. form, image
eluthero .. true
em, en .. in, into
encaphalo .. brain
endo, ento .. within
eo .. dawn ephippi .. a saddle
epi .. upon, to, cover

cardia .. heart
carpic .. wrist
carry-o, karya .. nut, nucleus
cata .. down, downward
ceno .. recent, common
cephal - e .. head
cera, cerato, kera .. horn
cero-o .. tail
cheeta .. bristle
chan-o .. gape, oven
chao .. empty space
chela .. claw
chilli, chilo .. thousand
chilo .. lip
chiro .. hand
chic - a .. a blade of grass
chloro .. green
choano .. funell
choric .. skin, membrane
chromo .. time
chryso .. gold
chthon .. land
cirrho-o .. orange coloured, twany
clade .. branch, sprout
cleido .. key, clavicle
cloisto .. closed
cleithro .. bar, bolt, key
clino .. bed
clipeus, cly .. shield
cool-o .. hidden
cteno .. comb
cyanc .. dark blue
dactylo .. finger
dasi, dasy .. hairy
del, delo .. visible
dondro .. tree
dar-o, dir .. neck, old, hide
derm-o .. skin
di .. double, two, twin
dia .. through, across
diadrom .. wandering
dino .. terrible, whirling
dolicho .. long
dropen .. sickle
hypso .. high
lchthy-o .. fish
infundibulum .. funell
leo .. equal
leio .. smooth
lepido, lepis .. scale
lepto .. fine, slender
leuce, leuko .. white
limno, limne .. lake
litho .. rock, stone
littor .. seashore
lopho .. crest
lute, luti .. yellow
erythro .. red
eso (oeso) .. within
etes .. one who dwells
ethmo .. a sieve

etymo .. truth
eu .. gocoo, wall, true
eum .. the place where
eury .. wide, broad
flab .. yellow
gastro, gastero .. stomach
glauco .. grey
glossa, glotto .. stomach
glypno, glypto .. engraved
gnatho .. jaw
gompho .. angles
gonic .. angled
gymno .. naked
gyno .. female
gyro .. round
hali, halo .. sea, salt
halpo .. single, simple
helec, helo .. marsh
helio .. sum
hemato, haemo .. blood
hepato .. lives
hetero .. different, other
hippo .. horse
hispid .. hairy
holo .. whole
homo .. like, same
hormon .. excite
hyalo .. glass, transparent
hydro .. water
hygro .. moist
hyo .. 'u'/'y'-shaped
hyper .. over, above, excessive
hypo .. under, beneath
pan, pant-a .. all
para .. beside, by the side of
pelagos .. sea
pelt .. shield
peri .. around, about
perisso .. out numbered
phago .. to eat
pheo .. dusky grey
phora .. to carry, bear
photo .. light
phyllos .. leaf
physa, physo .. bladder
phyto .. plant
plagio .. sideways
platy .. flat
pleco, pleoto .. twisted
pleo .. more, swis
pleuro .. lateral
pliesto .. most
plio .. more
ploto .. float, sail
pneumo .. lung
podo, pes, pug .. foot
poecili .. many colored
macro .. large, long
malaco .. soft
mastig-o .. whig
mega, megalo .. large
mers .. part, thigh

mees, mes .. middle
meta, met .. after, behind
micro .. small
mio .. less
monime .. fixed
mono .. single
myo .. muscle, mouse
myso .. muck
manne .. dwarf
noo .. new
neoto .. swim
nephre .. kidney
neuro .. nerve
nigro .. black
noto .. back
ochr .. pule yello
odenta .. tooth
oiko .. house
oligo .. few, less
onco, onono .. barb, hook
onycho .. claw
operculu .. cover, lid
ophi-o .. snake
opistho .. behind
opthalmo, opte .. eye
ornitho .. bird
ortho .. straight
ost, ostop .. bone
oto .. year
oxy .. sharp, pointed
pachy .. thick

palso .. ancient
scapho .. host
scat-o .. dung
schiate, ochizo .. split, cleave
scler-o .. hard
scolux .. word
scelio . crooked
scypho .. cup
saleni .. moon
sestro, sesto .. sieve
soleno .. pipe, channel
soma-to .. body
sphee .. wadge
sphyro .. hammer
splanchno .. viacera
spondylo .. vertebra
stage .. cover, roof
steno .. narrow
steonano .. crown
stetho .. cheat
stinums .. spot, brand
stema-to .. mouth
s[illegible]t[illegible] .. twiword
syn, sys .. together, with
pogono .. beard
poly .. many
porphyr .. purple
potano .. river
pro .. before
procto .. anus
profund .. deep

proto .. first
psamno .. sand
psoudo .. false
pters .. wing
ptilo, pterulo .. feather
pulchellus .. beautiful
pulmo .. lungs
pyrrho .. orangeeree
rach, rhaci .. spine
rumi, ramus .. branch
ramph .. curved bork or hook
rhin, rhino .. shark
rhino .. none
rhiso .. root
rhado .. rogs, red
rhyncho .. beak
sapro .. rottan
sarco .. flesh
sauro .. lizard

tule, to-o .. and, for, distant
telos, complete
thermos .. heat
topca .. rough
trachea .. windpipe
tracholo .. neck, throat
triono .. hair
troono .. wheel
tropho .. food, nourish
tropo .. twin
tycha .. chance
uro, ours .. twill
xantho .. yellow
xero .. dry
xeno .. stranger
xipno, zipho .. sword
xylo .. wood
xygo .. yoke, write.

Introduction

The 'Preface' depicts very briefly tit bit of initial success in research, some of which were nevertheless outstanding. More of such success recorded mostly in later years are now presented to testify to the Professor's vast experience that lay behind drawal of the outline of the three 'Manuals' now being edited.

Professor Chaudhuri was invited by FAO/UNDP in 1966 to develop Fish Culture, Fish Breeding and Training of Fishery Officers in Burma (now Mynamer). His outstanding success made FAO/ UNDP to extend the deputation period from three months to 4 years. In 1968 he was invited by FAO/UNDP to deliver lectures to Seminars on "*Genetic Selection and Hybridization of Cultivated Fish*" held in USSR for one month. He resigned from the services of Government of India in 1976 as he felt that under the existing system his vision of research, teaching and training would not fructify if he continued to serve there.

Immediately after his resignation from the Central Government services Professor Chaudhuri got an invitation to join SEAFDEC (South-East Asian Fisheries Development Center) Aquaculture Department as the *Regional Aquaculture Coordinator* to monitor the Aquaculture Programs of the South-East Asian countries. Also currently held the post of *Deputy Director of Aquaculture* when Asian Institute of Aquaculture was initiated at SEAFDEC.

In 1979 he joined FAO/UNDP as the Project Coordinator/ Project Manager/Chief Technical Adviser of the Aquaculture Development Project at Vientiane, Lao, PDR. Retired from FAO of the United Nations after serving for over nine years in the organization.

In 1984 he was invited to join SEAFDEC Aquaculture Department as its *Senior Technical Adviser* at Tigbauan, Iloilo, Philippines. Also, held simultaneously the post of NACA Training Coordinator under FAO and SEAFDEC/UPV training program for Senior Aquaculturists of Asia and the Pacific for one year (1986-87) leading to the degree of Master of Science in Aquaculture.

In 1988 he was offered Visting Professor's post in the Institute of Biological Sciences, College of Arts and Sciences, University of the Philippines at Los Banos.

Resigned from UPLB in March, 1993 to return to Calcutta, India and lead a completely retired life after fifty years of active professional career.

The editor is confident that the readers and users of the manuals would be able to fathom the depth of the Professor's knowledge from the invitations he received for accepting various assignments, and recognitions he got from some International Organizations only a few of which are mentioned hereinafter.

International recognitions and foreign assignments :

- Professor Chaudhuri was offered an assignment by FAO in 1964 to advise the Govrnment of Taiwan on induced breeding of Chinese carps in recognition of his pioneering work in breeding Asiatic carps by fish pituitary injections.
- Was selected by the organizers of the FAO World Symposium on Warmwater Pondfish culture to write a review on the *Breeding and Selection of Pondfishes of Asia and the Far East* and was also invited to act as the *Discussion Leader* for the session on Breeding and Selection of Warmwater Pondfishes at the Symposium.
- Was offered a second FAO assignment to help the Government of Burma in the development of fish culture and fish breeding techniques. The three months' assignments was successfully completed in 1966 (Rep. FAO/UNDP TA 2298).
- Was praised further as valuable contributions were made towards overall development of Fisheries in Burma with special reference to their fish seed requirements by induced breeding of carps; by successfully introducing Chinese carps and demostrating high production through

polyculture and training of over twenty Fisheries workers (Rep. FAO/UNDP TA 2954).

- Was also selected by FAO for delivering 3 lectures and taking active part in the FAO/UNDP Regional Seminar on *Induced Breeding of Cultivated Fishes* held at Cuttack, Bombay and Barrackpore during 1969.
- Was offered the 4th assignment by FAO for working as an Aquaculture expert initially for a period of three years at Bogota, Colombia in 1972.
- Was offered the 5th assignment by FAO in 1974 to work on the challenging problem of breeding milkfish in Indonesia.
- Was set on deputation by the Government of India in 1974 to advise the Government of Fiji in their *Grass Carp Project.*
- As FAO Consultant and Member, attended the Indo-Pacific Fishery Council Meeting of the *Working Party on Aquaculture and Environment* at Jakarta, Indonesia during October-November, 1974.
- Was appointed as a Consultant by the International Development Research Centre (IDRC), Canada for one month to assist the researchers in their Carps (Malaysia) Project at the Freshwater Fisheries Research Station (MARDI) at Batu Berendum, Malacca, Malaysia.
- Was invited to contribute a paper at the 13th Pacific Science Congress held at Vancouver, B. C., Canada; Two Papers were read and discussed.
- Organized and implemented the *Rural Aquaculture Project* in India, financed by IDRC, Canada as its Project Leader/ Co-ordinator.
- Was invited by FAO to act as a Panel member for the session *Pond culture of Finfish* at the FAO World Conference on Aquaculture at Kyoto, Japan (1976) and to contribute a paper on 'Problems of warmwater fishseed production'.
- Was offered the post of the *Regional Aquaculture Coordinator* by SEAFDEC a Regional Organization with Japan and several South-east Asian Countries as members, in recognition of his outstanding contributions to aquaculture research and development. Served SEAFDEC Aquaculture

Department from April, 1976 to November, 1979.

- Was invited by ESCAP (Mekong Committe) in February, 1977 to join a mission to Lao PDR as an Aquaculturist to make a feasibility study of a 50 ha fish farm which assignment was successfully completed.
- Was offered the post of Project Manager for the above project by ESCAP, Bangkok.
- Worked as a Consultant of IDRC, Canada at their Aquaculture (Sudan) Project in Khartoum, Sudan and also at the Carps (Malaysia) Project in Malacca, Malaysia in 1978 to assist and advise the Project Leaders in developing a specific research program and monitor other technical aspects of the projects.
- Was offered by World Bank in 1978 to join a pre-appraisal World Bank Mission to Banglandesh to evaluate financing in Inland Fishery Project.
- Was offered a Consultancy visit to the IDRC Inland Fisheries (Sarawak) Project as adviser for breeding fishes of commercial importance in Sarawak, Eastern Malaysia.
- Was offered by FAO to help their NACA (Network of Aquaculture in Asia and the Pacific) Project (in collaboration with SEAFDEC/UPV) as the NACA Training Co-ordinator, to organize one year's Training Course and deliver regular lectures to a group (22) of Senior Aquaculturists of Asia for their degree in M.S. in Aquaculture. Took all responsibility for the 1st year but declined the offer for the 2nd year as he had already accepted the invitation of Visiting Professor's position from UPLB.
- Was offered again a short term consultancy by FAO/UNDP to organize in Lao PDR, basic facilities for Applied Research and future planning of Applied Research and in-depth studies on Integrated Farming Systems in Lao, PDR, which consultancy was succussfully completed in 1986.
- Acted as a Consultant to the Philippine Council for Aquaculture and Marine Research and Development (PCAMRD) for 3 months in 1988 at Los Banos.

- Was offered a Consultancy Visit by International Development Research Centre (IDRC) to advise the Government of Malawi in East Africa to help Aquaculture projects in their Fishfarms.
- As a part-time Senior Aquaculture Consultant to International Aquaculture Consultants, made a Consultancy visit to Ellah Lakes Limited Fishfarm Project at Obrikom, near Port harcourt, Nigeria (West Africa) in 1988 as a Technical Adviser to advise on the overall development of the huge Fishfarm Project.

[The editor considers the following as a permanent testament to the professor's profused authority that speak on Aquaculture, Limnology and Ichthyology].

Rare International Honour for Dr. Hiralal Chaudhuri, Indian Fishery Scientist

[Source : Fishing Chimes, vol - 19, Nos. 10 and 11 Jan-Feb 2000]

Dr. Hiralal Chaudhuri, the eminent Fisheries Scientist, was recently honoured in U.S.A. for his outstanding contributions in the field of finfish aquaculture. In September 1994 a colloquium was held at the Bodega Bay Marine Laboratory in California, U.S.A., on the 'Applications of Endocrinology to Pacific Rim Aquaculture'. It was organized by the Universities of California and Tokyo with particular efforts made by Dr. Howard Bern, world renowned endocrinologist. Thirty two distinguished endocrinologists/ aquaculturists of the Pacific Rim countries were invited to participate and present papers. The plaque presented to Dr. Chaudhuri read 'Colloquium of the Bodega Marine Laboratory of the University of California, Davis, on Applications of Endocrinology to Pacific Rim Aquaculture' is pleased to honour Professor Hiralal Chaudhuri by dedicating its proceedings to him in recognition of his pioneering efforts in the use of reproductive hormones in finfish aquaculture and for training of generations of students in India, Laos and the Philippines. We commend him for his sustained international contributions over five decades and for his talent in applying the results of basic research to the solution of practical problems. The papers presented at this seminar are to be published in a special volume of the renowned journal 'Aquaculture' and dedicated to Dr. H. Chaudhuri.

This well-deserved rare honour to Dr. Chaudhuri is the crowning glory of an outstanding scientist whose contributions have made possible the planned fresh water aquaculture of today in respect of induced breeding of carps by the administration of fish pituitary hormones, the most important scientific methodology developed in the field of culture fisheries. This pioneering research work done by him in the fifities made possible to obtain seed of good quality, instead of the earlier method of collecting mixed seed (desirable and undesirable fish species together) from natural sources. His work possible the raising of unmixed fish seed in farm ponds, and subsequent stocking of waters with the kind of fish seed desired.

Dr. Chaudhuri did not rest with the development of this scientific method of breeding fish but carried on to take this to its logical conclusion i.e., optimizing production from fish ponds. He developed high yielding carp culture techniques in the late sixties, the first outstanding results obtained in Burma when serving as a FAO/UNDP Inland Fishery Biologist, he achieved high production (10 tonnes/ha) of carps in fish ponds. Later, on return to India he demonstrated the practicability of achieving high fish yields in ponds by carp polyculture (composite fish culture) through applied research work in Cuttack at the Central Inland Fisheries Research Institute, the same Institute where he had developed the induced fish breeding technique.

The excellent instructional film 'Induced Fish Breeding' produced by the Films Division of the Government of India was made at Cuttack under Dr. Chaudhuri's technical guidance. The film had received national honour as also the award for the best film at the FAO sponsored symosium on Aquaculture in Kyoto, Japan in 1976.

Dr. Chaudhuri combines in him the rare abilities of an excellent teacher the profession he started with after his brilliant academic career at the University of Calcutta as the outstanding researcher and the highly successful extension scientist. After joining the newly formed Central Inland Fisheries Research Institue in the late forties, Dr. Chaudhuri was sent in the mid fifties on deputation to the U.S.A., to work under Fish and Wildlife Service, by the Government of India. There he obtained the M.S. Degree of the Alabama Polytechnique working under the guidance of the Guru of Fish culture Prof. H. S. Swingle. The Institute is presently the Department ot Fisheries and Allied Aquacultures, Auburn University. It continues to be most famous fisheries school in the world.

Dr. Chaudhuri, as would be expected, has been showered with accolades and awards. He received the Chandrakala Hora memorial gold medal from the National Institute of Science in India in 1962, mainly for succeeding (first time in the world) in breeding Asiatic carps and developing a viable technique for large-scale production of their seed. He is also the recipient of the prestigious Rafi Ahamed Kidwai Memorial Gold Medal and prize awarded in recognition of his valuable contributions to the developments of aquaculture in India resulting in very high yield of fish protien.

In recognition of his outstanding contribution in the field of fisheries and aquaculture he was offered many consultancies/ assignments by the FAO, IDRC, ESCAP, WORLD BANK, SEARCA and SEAFDEC to work for the development of fisheries and aquaculture and training personnel in a number of developing countries of the world.

He served the FAO/UNDP as a fisheries expert for about 10 years working in Burma and Lao PDR. He was the Regional Aquaculture coordinator of South East Asian Fisheries Development Centre (SEAFDEC), Philippines, as also it's Senior Technical Adviser and visiting scientist for about 7 years and made significant contributions towards the success in breading of the milkfish, Chanos, besides training a host of visiting fisheries officials from different countries.

A unique appreciation came his way in 1994 when the University of Calcutta, his alma mater, instituted a chair in Fisheries in Chaudhuri's name.

Inland aquaculture owes a permanent debt to Dr. Hiralal Chaudhuri for making possible through his painstaking brilliant efforts production of quality fish seed and how best to culture them through scientific farming to obtain high yields of food fish.

A special issue of the journal

"World Aquaculture" (1995) Reads-

"This special issue is comprised of the papers and abstracts presented a colloquium held in honour of Professor Hiralal

Chaudhuri at the University of California Bodega Marine Laboratory on 07-11 Sept. 1994. This was the seventh in a series of annual conferences held at the Laboratory.

This topic was chosen because all Pacific Rim countries have an increasing interest in the practise of aquaculture to supplement food production. Owing to intense fishing pressures, the aquatic culture of shellfish and finfish plays and will continue to play a major role in the economics and research thrusts in most of these Pacific Rim Countries. One area of aquaculture research that is currently experiencing intense effort is hormone manipulation. Because we had to keep the meeting small and focused we were constrained to limit North American participation in the meeting to colleagues in the western provinces and states of Canada and the USA.

The papers were divided into section on reproduction, development, salmons moltification, growth, metabolism and general endocrinology. There was also an evening workshop on maturation and reproduction in captive salmonid populations. The meeting stressed a comparative approach and included papers on invertebrate and vertebrate species of interest of aquaculture.

Professor Hiralal Chaudhuri, the participating guest of honour at this meeting, receive our grateful recognition for his pioneer efforts in the use of reproductive hormones in finish aquaculture and for his training of generations of students in India, Laos, Myanmar (Burma) and the Philippines. Professor Chaudhuri received his doctoral degree in Zoology from the University of Calcutta in 1961 and has provided counsel to the aquaculture programs of many countries, including India, the Philippines, Laos, Myanmar, the former Soviet Union, Indonesia, Malaysia, Thailand and Fiji under the auspices of the Food and Agriculture Organization of the United Nations. He is truly an international figure. Professor Chaudhuri has been called the "Father of Induced Breeding" his pioneer study on the use of a pituitary preparation to induce spawning in carps was published in 1957."

These high grade assignments and sky-high recognitions in international arena clearly testify to Professor Chaudhuri's wide range of knowledge over several subjects such as Botany, Zoology (all branches), Ecology, Chemistry, Biometry besides his knowledge of several languages (English, French, German and Latin). Such wide knowledge mingled with affectionate approach, patient deliberations all with a sacred missionary zeal made him a very successful teacher and a trainer.

The students, both under and post graduate, of the Los Banos University of the Philippines gave this Professor their highest "evaluation value' for the best teaching quality among all the teachers. A system was vogue in the University by putting value (mark) on various aspects of teaching.

Not only with University, he also impressed generations of trainees in various fields of aquaculture during his service tenure with Government of India (prior to 1976) and thereafter in International Organizations (from 1976 to 1993) which made him all time a successful trainer.

Professor's Unique Strategy in Teaching

Besides his deliberation by the aid of colour slides, transparencies, herberium sheets, preserved specimens, hand drawing and free hand documentations the professor used to break monotomy of his lecture by presenting several of his research findings of his technical work on the following subjects in the recess hours :

- Embryonic development of the transparent Goby, *Goblopteus Chuno.*
- Effect of pituitary injections and other factors on the reproduction of pond fishes.
- Experiments on hybridization of Indian carps.
- Experiments on induced spawning of Indian carps with pituitary injections.
- Induced spawning in Indian carps.
- Induced breeding and development of common Catfish "*Ompok bimaculatus*".
- Spawning and hybridization of Indian Carps.
- Induced spawning of cultivated fishes.
- Techniques of hypohysation in Indian Carps.
- Methods adopted for inducing breeding of fishes.
- Breeding habits of cultivated fish.

- Stocking pond : management in composite fish culture.
- Culture of air-breathing fish.
- Use of hormones in induce spawing of Carps.
- Experiments on large-scale productions of seed of the Chinese carps.
- Use of hormones in breeding cultivated warm water fishes with special reference to milkfish *Chanos chanos.*
- Proteins of warm water fish seed production.
- Method demonstration to educate fish culturists.
- Induce Breeding of Carps.
- Induce Breeding of Asiatic Carps.
- Studies on the effect of grass carp on paddy plankton.
- Notes on external sex characters of *Chanos chanos.*
- History of Induce Breeding in fishes and its application to aquaculture.

With all these background information the reader may find manuals very educative.

PART ONE

AQUACULTURE

1
PREAMBLE – DEFINITION

Definition

"Culture of any aquatic organism of economic value."

The organism may vary from

The Finest Unicellular Algae to the Giant Seaweeds or from microscopic protozoa to a big fish or even a huge mammal such as a whale.

Aquaculture is the aquatic counterpart of argriculture – an all-embracing team including culture of any aquatic organism.

The aquatic organism may vary from the finest unicellular algae to the giant seaweeds or may vary from the minutest microscopic protozoan to a big fish or even a huge aquatic mammal like a whale.

Other definitions of aquaculture

Aquaculture is the rearing of aquatic organisms under controlled or semi-controlled conditions (a simple definition).

From the biological point of view

Aquaculture may be regarded as man's attempt through inputs of labour and energy to improve the yield of useful aquatic organisms by deliberate manipulations of their rates of growth, survival and reproduction.

Yet there is another definition

The term 'aquaculture' is coined to embrace all the activities of man to produce through husbandry during the whole or part of the life-cycle of any aquatic resource of economic value.

Aquaculture has been identified as one of the major strategies to accelerate world fisheries production.

2
AQUACULTURE

A Major Strategy to Enhance World Fish Production

Technological advances made in recent years through :

- Polyculture (multigrade/mixed size and age stocking)
- Cage and pen culture/culture in enclosure
- Running water culture/recirculating tank system
- Integrated fish/livestock/crop production systems
- Recycling of wastes/sewage-fed fish culture
- Sea-ranching/artificial recruitment/transplantation
- Controlled reproduction
- Improving strains of fish by genetic selection and hybridization.
- Development of high protein balanced feeds
- Introduction of efficient exotic species
- Intensive research/technology generation and dissemination
- Development of technical manpower
- Development of institutions
- Development of infrastructures
- Massive financial investments
- Suitable legislations.

History of Aquaculture Development in Various Countries of the World

Aquaculture originated and developed under different social and economic conditions in different countries.

China : Aquaculture was undertaken to assure a reliable supply of an essential item of food.

India : Areas excavated for construction (flooring) of houses were later on used for fish culture (freshwater ponds) brackish water ponds developed as a stage in land reclamation – to raise land level for agricultural purposes.

Thailand : Originated as a multiple use of salt pens (also probably in the Philippines).

Indonesia : The "tambak" system was started by convicts in 13th century who were forbidden to practise agriculture or other professions.

Europe : Aquaculture began in medieval times mainly in association with monasteries.

Africa : Initiated on a subsistence level by colonial administration by improve nutrition level of natives. (Relatively recent).

North America : With an emphasis on the production of sport fish (bluegill and bass).

"But inspite of such clear differences in patterns of development, present-day interest in aquaculture has a common orientation towards the establishment of a viable industry irrespective of social and economic differences." (Table 1 and 2)

Table 1 : Summary of Aquaculture Production by Region : 1983

(in metric tons)

Region	*Finfish*	*Molluscs*	*Crustaceans*	*Seaweed*	*Total*	*%*
Africa	43,553	286	26	–	43,865	0.4
Asia and the Pacific	3,357,978	2,586,464	75,644	2,392,045	8,412,131	82.4
Europe and Near East	726,530	494,719	162	100	1,221,511	12.0
Latin American and Caribbean region	167,797	30,883	20,108	1,697	220,505	2.2
North America	152,088	133,178	27,425	–	312,691	3.0
Total	4,447,946	3,245,530	123,445	2,393,782	10,210,703	100.00
Percentage (%)	43.56	31.79	1.21	23.44	100.00	

Table 2 : Fish Production in the Philippines by Aquaculture (1986) (vol. 36, 1986 BFAR)

Sl. No.	*Type of Aquaculture & Product*	*Quantity (mt)*	*Value (P 1,000)*
1.	Marine	197,447	527,187
	Oysters	16,465	85,618
	Mussels	12,114	107,210
	Seaweeds	168,868	334,359
2.	Freshwater	50,669	1,257,667
	From fish pens	41,296	1,035,027
	From cages	9,373	222,650
3.	Fishponds	222,777	9,047,337
	Brackish-water	207,274	8,682,201
	Freshwater	15,503	365,135
	Total	**470,893**	**10,832,201**

Total Fish Production in the Philippines (1982-86)

Year	*Production (mt.)*	
1982	1,896,983	(Base year : 1978-
1983	2,110,230	1,580,404 mt.)
1984	2,080,439	
1985	2,052,111	
1986	2,089,484	

Aquaculture is forecast to displace the fuel-dependent and calamity prone marine fisheries sector in the Philippines as a result of massive shifting to fish farming.

3

Aims and Objectives of Aquaculture

The aim of aquaculture as popularly envisaged is to grow desirable species of fish to sufficiently large size within a minimum period of time thereby harvesting maximum quantity of fish from a unit body of water.

Objectives of Aquaculture

- Human food production: a good source of protein
- Production of sport fish: for recreation
- Production of bait fish:
  a. sport fishing
  b. commercial fishing
- Improvement of natural stock: transplantation/artificial recruitment
- Production of ornamental fishes
- Production for the biological control of:
  a. mosquitoes by larvicidal fishes
  b. aquatic weeds by macrophytophagus fishes
  c. snails by malacophagus fishes
- production of natural feeds for larval rearing in hatcheries
- production for industrial purposes:
  a. pearl culture
  b. culture of seaweeds for colloids.

4

Basic Principles Underlying Fish Production

1. Animals are dependent upon plants for food.
2. Fish production on pond depends on its capacity to produce plants.
3. Plant production depends on: sunshine, water, CO_2 and essential elements: C, H, O, N, P, K, S, Ca, Mg, B, Zn, Si and Cu.
4. Fertility of watershed limits productivity of natural waters.
5. Productivity of water can be increased by fertilization.
6. Phosphorus is the first limiting factor.
7. Next limiting factor is nitrogen.
8. Carbon dioxide is often the next limiting factor.
9. Demand of oxygen both by living and dead organisms is the next limiting factor.
10. Of plants, phyto-plankton appears most desirable for their:
    - short life span
    - small size
    - high nutritive value
    - greater surface area
    - they do not interfere with movement of fish.
11. A. higher aquatic plants (rooted) desirable to limited extent:
    - Absorbs nutrients from pond bottom
    - Release them into pond water on their death
    - Also from food of fish

B. Floating plants are undesirable because they:

- cut off sunlight
- low production of fish-food organisms
- low oxygen content of water

12. Shorter the food chain higher the production.
13. At a certain level of fertility, a pond can produce only a certain poundage of fish.
14. The poundage may be a large number of small fish or a small number of large fish.
15. Large number of small fish can utilize the available food more fully than a smaller number of large fish.
16. For a short period of time the number of fish can be regulated by the stocking rate.
17. For longer period of time the number can be regulated by using a piscivorous fish.
18. A balance between the population of piscivorous species and forage species to be maintained.
19. The greatest weight of a given species can be produced by stocking only that species.
20. Maximum production, however, is obtained by using combination of species of different feeding habits.
21. It requires more food to grow a pound of fish than to maintain it.

WHY CULTURE FISH?

- A good source of cheap protein
- Easily digestible
- No fear of blood cholesterol
- Gives higher production than livestock
- Growing fish is cheaper than poultry/piggery
- Can utilize all available food in a pond
- Grows faster
- Has an efficient FCR (food conversion ratio)

FOOD INTAKE IS SPENT ON

Maintenance

- for energy
- for movement
- for thermo-regulation
- for reproduction.

Growth

(most important for fish farmer)

Food used for maintenance is a waste for fish farmers.

Fish spends less energy for maintenance. why?

- It does not waste energy in carrying the weight of its body.
- It does not waste energy in heat regulation

PER UNIT OF ASSIMILATED FOOD

- 1.5 times more than poultry/pig
- 2.0 times more than cattle/sheep

Why? Because of efficient FCR.

5

Primary Production and Food Chain

Primary production:	Is the energy accumulated by plants in photosynthesis
Primary productivity:	It is the rate of energy storage (per unit of time) by plants.
Cross productivity:	Is the total rate of energy storage.
Net primary productivity:	It is determined by subtracting energy consumption by respiration of plants from the gross productivity.
Biomass:	Biomass is expressed as the weight of dry organic matter per unit area and differs from productivity, which is the rate at which organic matter is formed by photosynthesis.
Food chains:	Are descriptions of the energy flows through the ecosystem. Food chains follow a pattern. Primary producer (plants), primary consumer (herbivore), secondary consumer (carnivore), tertiary carnivore quaternary carnivore-top carnivore (man). (Table 3)

Table 3 : Ecological Pyramid

A. Pyramid of Numbers	*B. Pyramid of Mass*	*C. Pyramid of Energy*
Second level carnivore 10	Owl: 5 tons Skunk 1.5 tons	Man 8.3 x 10^3 cal
Carnivores first level 90,000	Mice 100 tons	Beef 1.16 x 10^6 cal
Herbivores 200,000	Grasshopper 900 tons	ALFALFA 1.49 X 10^7 cal
Producers 2,000,000	Grass 5,000 tons	
This contains no information about size or energy content of each level in the food chain	It gives better idea of the energy requirement. But some times misleading when organisms having lot of mass contribute very little as food material.	It best illustrates the actual energy relationships within the community organisms. But often very difficult to determine the calorie (energy) values.

"If men are to derive greater benefit from the oceans as the food source, they must eat more of the fishes that are lower in the food chain." (Table 4)

Table 4 : Culture of Penaeus Monodon : Grow-Out Systems

Intensity of culture	*Size of Pond*	*Depth*	*Fertilization*	*Supplementary Feeding*	*Water Management*	*Stocking Rate*	*Production (per ha)*
Extensive (Traditional and improved traditional)	3-20 ha	0.3 - 0.6 m. Mx 1 m in trench	Yes Both inorganic and organic	No	Hardly any (mainly tidal)	3,000-5,000 Fry/ha. Less than $1/m^2$ or $1\text{-}2/m^2$	50-500 kg/ha
Semi-extensive (Much improved traditional)	1-3 ha	1.8 - 1.2 Trench	Yes Both organic and inorganic	Yes Trash fish and formulated feeds	Regular Renewal of water with water pump 5-15%	20,000-50,000 fry/ha $5\text{-}15/m^2$	500-1000 kg/ha 1-6 ton/ha (Amakasu type in Japan)
Intensive (Sophisticated)	0.05-0.5 ha (earthen or concrete circular tanks)	2 m	Yes Mostly synthetic	Yes Intensive feeding with formulated high protein diet	Intensive management Renewal of water : 90-100% Continuous elimination of metabolites	$20\text{-}50/m^2$ $100\text{-}250/m^2$	5-8 tons 10-40 tons (Shigueno type)
Superintensive	- do -	- do -	- do-	- do -	High water exchange	$500\text{-}600/m^2$	USA

6
CLASSIFICATION OF AQUACULTURE

A. According to purpose:

- Food production
- Sport fishing
- Bait for commercial fishing
- Pearl/colloids industry
- Ornamental fish
- Improve natural stock
- Recycling of organic waste
- Integrated farming

B. According to culture environment:

- Ponds/tanks
- Cages/pens/enclosures/raceways
- Coastal water/estuaries/lagoons/open sea
- Silos/compact recirculating tanks

C. According to intensity of culture:

- Traditional
- Extensive
- Semi-intensive
- Intensive

Constraints

- Technological (Generation and transfer)
- Financial (Credit)
- Man-power (Need for training)

- Environmental (Aquatic pollution)
- Socio-economical (Pouching/use of pesticides)

HIGH COST OF SEED - FEED - LABOUR.

Choice of Species for Aquaculture

Desirable characters

- Quicker growth
- Short food chain
- Efficient food conversion ratio (FCR)
- Hardy: capable to withstand sudden change of environment
- Meet consumers preference (high table quality)
- Efficient in utilizing natural food
- Readily accepts compounded feeds
- High fecundity
- Early maturation
- Easy to breed in captivity
- Prolonged breeding period
- Compatible to other species
- Disease resistant
- Tolerance to wide range of environmental parameters: temp., pH, salinity, DO^2
- No adverse environmental impacts
- Ability to tolerate crowding
- Grow to fairly large size
- Easy to handle.

Criteria for selection of species for aquaculture

- Objectives of culture
- Geographic and climatic considerations
- Culture qualities of organisms
- Consumer acceptance and marketability (quality product/ consumer preference/market price/marketing strategy)

- Cost of production (Cost of culture units/cost of inputs/ cost-of management/taxes/loan amortization/others)
- Domestic consumption vs export (Low priced species/high priced species).

Aquaculture of marine finfish

Important cultivated species:

Scientific Name	*Common name*
Seriola quinqueradiata	Yellow tail
Epinephelus tauvina	Grouper
Epinephelus salmoides	Grouper
Eugu rubrifies	Puffer
Eugu vermiculatus	Puffer
Pagrus major	Red porgy
Mylio macrocephalus	Black porgy
Pleuronectes platessa	Plaice
Solea solsea	Sole
Scophthalmus maximus	Turbot
Pseudopleuronectes americanus	Flounder
Microstomus kitt	Lemon sole
Trachnotus carolinus	Pompano
Salmo salar	Atlantic salmon
Oncorhynchus spp.	Pacific salmon
Euryhaline Fishes	
Chanos chanos	Milkfish
Mugil cephalus	Grey mullet
Lates calcarifer	Sea bass
Etroplus suratensis	Pearl spot
Siganus guttatus	Rabbit fish

BRACKISHWATER POND CULTURE
IMPORTANT CULTIVABLE SPECIES

A. Finfish:

Scientific Name	*Common name*
Chanos chanos	Milkfish
Mugil cephalus	Grey mullet
Mugil parsia	Mullet
Mugil dussumieri	Mullet
Liza macrolepis	Mullet
Liza tade	Mullet
Lates calcarifer	Seabass
Etroplus suratensis	Pearl-spot
Siganus guttatus	Rabbit fish
Epinephelus spp.	Grouper
Oreochromis mossambicus	Java tilapia
B. Crustaceans:	
(a) Shrimps and prawns:	
Penaeus monodon	Tiger shrimp
P. japonicus	Kuruma shrimp
P. indicus	White shrimp
P. merguiensis	Banana shrimp
Metapenaeus ensis	Greasy back shrimp
M. monoceros	
M. dobsoni	
M. brevicornis	
P. orientalis	
P. semisulcatus	
P. latisulcatus	
(b) Crabs:	
Seylla serrata	Mud crab
Neptunus pelagicus	Blue crab

Aquaculture of freshwater finfish: Important species

Species (common name)	*Country where mainly cultured*
Fam: Cyprinidae (carp family)	
Cyprinus carpio (common carp)	All over the world
Hypophthalmichthvs molitrix (silver carp)	China/many countries
Ctenopharyngodon idella (grass carp)	China/many countries
Aristichthys nobilis (bighead carp)	China/many countries
Cirrhinus molitorella (mud carp)	China/many countries
Cirrhinus mrigala (mrigal)	Indian subcontinent & Myanmar
Catla catla (catla)	Indian subcontinent & Myanmar and many countries
Labeo rohita (rohu)	Indian subcontinent & Myanmar and many countries
Labeo calabasu (calbasu: black carp)	Indian subcontinent & Myanmar
Puntius gonionotus (tawes)	Indonesia/many countries
Mylopharyngodon piceus (black carp)	China/a few countries/ Indonesia Malaysia Thailand/Vietnam
Carassius auratus (goldfish)	Many countries
Carassius carassius (golden English carp) (golden/English carp)	
Parabramis pekinensis (bream)	Several countries/China

Species (common name)	Country where mainly cultured
Fam: Cichlidae (Tilapia family)	
Oreochromis niloticus (Nile tilapia)	Africa/many countries
Oreochromis mossambicus (Mossambique/Java tilapia)	Africa/many countries
Oreochromis aureus (blue tilapia)	Africa/Isreal/Taiwan
Tilapia zilli (zilli's tilapia)	Africa/Malaysia/ Philippines (exptal)
Etroplus suratensis (pearl - spot)	India/Sri Lanka
Fam: Centrarchidae (Carp sucker family)	
Ictiobus cyprinellus (bigmouth buffalo)	USA
Fam: Centrarchidae (Sunfish family)	
Micropterus salmoides (bigmouth bass)	USA
Lepomis macrochirus (Blue gill)	USA
Fam: Osphronemidae (Giant Gourami family)	
Osphronemus soramy (giant gourami)	Indonesia/Sri Lanka/ Malaysia/India
Fam: Belontidae (Gourami family)	
Trichogaster pectoralis (sepat siam)	Thailand/Malaysia/ Indonesia
Helostoma temmincki (kissing goramy)	Thailand/Malaysia/ Indonesia
Fam: Pangasidae (Catfish Family)	
Pangasius pangasius (butter fish/pangas)	India/Thailand/Laos

Species (common name)	Country where mainly cultured
Fam: Ictaluridae (American catish family)	
Ictalurus punctatus (channel catfish)	USA/a few other countries
Fam: Clariidae (Walking Catfish family)	
Clarias batrachus (magur/pladuk)	Indian subcontinent/South East/Asia Philippines
Clarias macrocephalus (hito) *Clarias gariepinus* (African catfish)	Egypt/Nigeria/Cent. Africa/many countries
Fam: Heteroponeustidae	
Heteropneustes fossilis (singhi)	Indian subcontinent
Fam: Notopteridae (feather back family)	
Notopterus chitala (chitala)	India/Bangladesh/ Thailand/Laos
Fam: Anabantidae	
Anabas testudineus (climbing perch)	Indian Subcontinent/ S.E. Asia/S. China
Fam: Centropomidae	
Lates calcarifer (seabass/ cock-up)	Thailand/India/ Philippines
Fam: Mugilidae (mullets)	
Mugil cephalus (grey mullet)	Many countries
Mugil parsia (parse)	India
Fam: Chanidae	
Chanos chanos (milkfish)	Philippines/Indonesia/ Taiwan/a few countries
Fam: Channidae or Ophicephalidae (murrels)	
Channa striata or *Ophicephalus striatus* (striped murrel)	Indian subcontinent Myanmar/S.E. Asia/ Philippines

Species (common name)	*Country where mainly cultured*
Fam: Anguillidae (eels)	
Anguilla anguilla (European eel)	Japan/Taiwan/Europe
Anguilla japonica (Japanese eel)	Japan/Taiwan/Korea
Fam: Salmonidae (trout and salmon)	
Salmo gairdneri (rainbow trout) / *Salmo trutta* (brown trout)	N. America/Europe/ Japan/many countries
Oncorhynchus keta (cherry salmon)/	USA/Canada/USSR/Japan
O. gorbuscha (pink salmon)	
PRAWN: Palaemonidae	
Macrobrachium rosenburghii (Giant freshwater prawn)	Many countries

7

Biological Means of Increasing Production

- Use of efficient species
- Use of species combination
- Determination of proper rate of stocking
- Stock manipulation/continuous harvesting
- Control of reproduction
- Other means of increasing production
    - o Cropping of relatively small fish
    - o Adoption of balanced forage/predator ratio
    - o Heavy stocking in running water with intensive feeding
    - o Aeration/elimination of metabolites
    - o Improvement of stock by hybridization.

What is Stock Manipulation?

To manage fish population in ponds until ready to harvest

Control of Reproduction

A. For fishes which do not reproduce in ponds;
- To induce spawning
    i. By environmental manipulation
    ii. Bys hormone injection

B. For fishes which breed prolifically in ponds:
- To prevent reproduction
    a. Mono-sex culture
    b. Sex reversal

c. Hybridization - all male hybrids

d. Artificial androe -/gynogenesis

- To control overpopulation

a. introduction of suitable predator

POLYCULTURE OF DIFFERENT SPECIES OF FISH

(Same as: mixed farming/association of species/mixed fish culture/composite fish culture)

It is an important management technique to utilize all food resources in a pond by a mixed population of ecologically different species in a most efficient way.

SELECTION OF SPECIES IS IMPORTANT

Criteria for selection:

- Quick growing
- Short food chain (herbivorous)
- Tolerant to other species (compatible)
- Having different feeding habits
- Feed in different ecological niches
- Have efficient FCR
- Utilize natural/cheap artificial food efficiently
- Able to reproduce in captivity
- Meet consumer's preference

Table 5 : Polyculture of Asiatic Carps

Popular Name / Scientific Name	*Important Food Items*	*Feeding Zone*
*Silver carp *(Hypophthalmichthys molitrix)*	Phytoplankton	Surface layer
* Bighead *(Aristichthys nobilis)*	Zooplankton	Surface layer
+ Catla *(Catla catla)*	Zooplankton	Surface layer
+ Rohu *(Labeo rohita)*	Periphyton / Decaying veg.	Middle layer (column feeder)

Contd...

Table 5 – *contd...*

Popular Name / Scientific Name	*Important Food Items*	*Feeding Zone*
* Grass carp *(Ctenopharyngodon idella)*	Aquatic vegetation	Surface / middle layer
** Tawes *(Puntius gonionotus)*	Aquatic vegetation	Surface / middle layer
* Mudcarp *(Cirrhinus molitorella)*	Bottom fauna / detritus	Bottom layer
+ Mrigal *(Cirrhinus mrigala)*	Bottom fauna / detritus	Bottom layer
* Black carp *(Mylopharyngodon piceus)*	Bottom organisms / molluscs	Bottom layer
Common carp *(Cyprinus carpio)*	Bottom organisms / omnivorous	Bottom layer

* Chinese carp

+ Indian carp

** Indonesian carp.

Fishes Used in Polyculture Other Than Carps

Java tilapia *(Oreochromis mossambicus)*	Omnivorous all layers
Nile tilapia *(Oreochromis niloticus)*	Omnivorous all layers
Sea bass *(Lates calcarifer)*	Piscivorous all layers
Chital *(Notopterus chitala)*	Small fish / insects / offal feeder all layers
Striped murrel (dalag) *(Ophicephalus striatus)*	Piscivorous all layers

EXTENT OF COMPETITION IN POLYCULTURE

Species 'A' is stocked with Species 'B'
(Their food habits apparently similar)

Rate of growth is expected to be same for A and B Stocking 1000A with 100B. B grows bigger than A again Stocking 100A with 1000B, A grows bigger than B. So, there is some difference in their food preference.

How to Determine the Rate of Competition?

Species 'D' is stocked with species 'E'

c = index of rate of competition

a = production of 'D' (by monoculture)

b = production of 'D' (polyculture with 'E')

Therefore, the formula is:

$$C = \frac{a-b}{a}$$

$= x$

Competition is more when x is high

When x is negative; there is positive influence of 'E' on the growth of 'D'.

FURTHER MODIFICATIONS OF POLYCULTURE SYSTEMS

Multiculture Stocking

It is founded on sound biological basis which calls for the maximum use of the pond space and the food available by using different species of fish (also their size and age group) with varying behaviour and feeding habits.

Polyculture by mixing various species with different sizes and age groups (as practised in Modern China).

1. Multigrade system: (Fishes are reared in a series of ponds from fingerlings to market-size) (Table 6).

This system takes advantage of the maximum growth potential of the fish as their density can be adjusted for the size of the fish as well as for the productive capacity of ponds. This enables the farmer to undertake a continuous cycle of stocking and harvesting, thereby increasing the output. The rearing period is shortened and production of the desired size of fish for market facilitated. (This technique, however, can only be practised in communes of China that have several units for large-scale production).

Table 6 : Stocking density of multigrade polyculture

Grade	*Size of fish (Wt. in g)*	*Species No. per hectare*				*Rearing period (No. of days)*
Species		Bighead	Grass carp	Silver carp	Mud carp	
Grade I	14-65	6,750	15,000	–	–	40
	14-80	–	–	3,750	–	
	5-20	–	–	–	112,500	
Grade II	65-225	2,100	6,750	–	–	150
	80-500	–	–	600	–	
	20-60	–	–	–	37,500	
Grade III	225-500	900	3,000	–	–	150
	500-1000	–	–	300	–	
	60-270	–	–	–	12,750	
Grade IV	500-1206	375	1,050	–	–	

2. Improved polyculture of mixed-age and mixed-size groups: (Multiple age method).

8
Distinctive Features of Fish Culture in the Various Regions of the World

DUE TO DIFFERENCE IN APPROACH TO SOCIOLOGICAL - CLIMATIC - BIOLOGICAL FACTORS

Far Eastern (Asian) System of Fish Culture

Small size ponds (0.3-2 ha); polyculture of 5-6 species; elaborate system of stocking and careful management resulting in high production with low expenses (use of animal and plant wastes). Here pond is a part of an ecological complex.

European Fish Culture System

Big fish farms (50-200 ha) producing mainly common carp (monoculture); sometimes with tench and pike. Ponds cropped once a year - growing market size in 3-year rotation mainly on natural food to avoid high expense.

Middle East (Israel) System of Fish Culture

Moderate size ponds (4-6 ha); very intensive management; stocking with two size carp with either tilapia or grey mullet.

Stocking policy: to adjust number of each fish groups to minimize competitions both intra and interspecific, full utilization of resources resulting in high production. (Yashouv, 1968)

WHAT IS SUBSISTENCE AQUACULTURE?

(Dictionary meaning)

Farming of a system or farming that produces a minimum and often inadequate return to the farm operator: economically marginal farming.

OR

Farming or a system of farming designed to provide all or essentially all the goods required by the farm family usually without any significant surplus for sale.

Character: (a) Low inputs

(b) Low level of technology (traditional methods)

(c) Low output

Yet others argue about the meaning of "Subsistence" (It is interpreted differently by different people).

"Subsistence would not be when a person is able to produce some fish for his own consumption, that would not be enough or adequate to afford subsistence for him. Actually he should be able to get an income that would ensure a standard of living comparable that of others in the society where he lives".

9

Subsistence Level Aquaculture is Sometimes Termed as : Homestead Aquaculture

Subsistence and small commercial scale aquaculture currently dominate aquaculture industry in many developing countries of the world.

VERTICALLY INTEGRATED FISH FARMING SYSTEM (VIFFS)

A vertically integrated aquaculture industry is one in which all different components of the farming system including hatchery management, grow out operations, are coordinated and controlled by centrally-managed systems. It is usually associated with large economic units, large production, large market and large profit.

Four Major Operations Integrated

A.	Supply of materials and services	Supply
B.	Farming itself (that is: breeding and growing)	Production
C.	Processing and	Processing
D.	Distribution	Marketing.

Vertical Integration gives scope for centralized planning. VIFFS is possible mainly in developed countries since the industry is a high-risk enterprise needing high capital investment. In U.S.A., VIFFS is favoured since such industry:

a. promotes research and large scale technology;

b. induces large capital investment;

c. utilizes team organization, and

d. assists production development and marketing.

Constraints

(i) Species limitations (market, biological and technological),
(ii) Sites
(iii) Feed
(iv) Manpower
(v) Legal and institutional
(vi) Financial.

VIFFS

Support system, Production system, Marketing system

It spreads profit or loss over different sections.

Merits and Demerits of the Two Farming Systems

Sl. No.	*Subsistence level fish farming*	*Vertically integrated fish farming*
1	Traditional (extensive)	Highly sophisticated (super-intensive)
2	Low inputs	High inputs
3	Low technology	High technology
4	Low outputs	High outputs
5	Low risk	High risk
	Age old practice in developing countries and still continuing	Suitable adapted in some developed countries.

A technologically developed developing country should take up (A) or (B)?

A via media path is preferable: Neither traditional nor too sophisticated. That is - Small scale commercial level aquaculture.

"Small Scale Commercial Level Aquaculture"

This type of culture is:

1. semi intensive
2. inputs not too high or too low
3. high technology
4. satisfactory output (profitable).

In majority of the developing countries, there will be a marked preference for small scale, labour-intensive form of aquaculture for integration with the rural development programme. Its success will largely depend on the support services that are available. The small farmers would greatly benefit from a package of technical assistance, provisions of inputs and institutional credits' on reasonable terms.

THREE-DIMENSIONAL AQUACULTURE

In this type of aquaculture the entire body of water from top to bottom is utilized for culture. The commodities are:

(A) Molluscs and (B) Seaweeds

(A) Molluscs

Cultured species are mostly the bivalves which are filter feeders such as, oysters (e.g. *Crassostrea gigas*), mussels (e.g. *Perna vifidis*), Clams, cockles and scallops.

These are cultured in coastal shallow waters (mostly in the intertidal zone) directly on the seabed or using stakes or poles, racks and trays. In deeper waters, floating rafts or long lines are used [culture by suspension].

Types of Mollusc Culture

1. Bottom Culture
2. Pole Culture
3. Tray Culture
4. Raft Culture
5. Long Line Culture.

(B) Seaweeds

Seaweeds, such as, Eucheuma, Gracilaria, Porphyra and others are cultured in protected Bays in shallow waters.

1. **Bottom culture:** Eucheuma, Gracilaria and Caulerpa are cultured in bottom habitats. The net or long line methods are used.
2. **Culture by suspension:** The red seaweeds Porphyra are cultured on rectangular (horizontally placed) nets tied to stakes. Long line method is used for Undaria and Laminaria.

SEAWEED CULTURE

Green Seaweed:	Caulerpa, Ulva, Enteromorpha.... Human food
Brown Seaweed:	Sargassm............ source of algin Undaria, Laminaria
Red Seaweed:	Porphyra, Gracilaria, Eucheuma, Hyupnea, Gelidiella, Gelidium.

In 1973: 2.4 million metric tons of seaweeds valued at US$765 million were produced. Bulk of these from Japan, China, Korea, U.S.A., Brazil, Norway and some other countries in Europe.

In 1980, Philippines exported over 13000 mt. of seaweed worth P 55.6 m.

Phyco-colloids from seaweeds

Agar: Prepared from - Gracilaria, Gelidium and Gelidiella; used for: As culture medium for bacteriological work; manufactures of gelatin; as anti-drying agent in bread and pastries; manufacture of frozen dairy products.

Algin: Prepared from Sargassm.

It is an important component in the manufacture of ice-cream and anti-biotics; used in rubber and textile industry.

Carrageenans: Prepared from Eucheuma and Hypnea.

Used in the manufacture of surgical jellies and ointment, cream, etc.

10
CRITICAL STANDING CROP (CSC)

Since production of natural food is limited, when a pond is stocked with a certain density of fish they will reach a weight above which the natural food will not be sufficient to sustain both maintenance and maximum potential growth rate. Because requirement of maintenance must be provided first, with the increase in fish weight over this point, the growth rate will be reduced. This is the critical standing crop (CSC) of the fish. Growth will cease entirely at the carrying capacity when available food is just sufficient to maintain the fish but insufficient to allow growth in order to maintain the potential individual growth rate beyond the CSC, the amount of food should be increased either by fertilization or by adding supplementary feed.

Food is not the only factor that affects CSC and MSC. As standing crop increases, the supply of oxygen may become insufficient. This will also happen as metabolites accumulate.

BOTH CSC AND MSC ARE EXPRESSED AS BIOMASS IN KG/HA.

Table 7 : For common carp in Israel (Hepher)

Sl. No.	*Feeding level*	*CSC (kg/ha)*	*Carrying capacity (kg/ha)*
1	No feed/no fertilization	65	130
2	No feed/fertilization	120	480
3	Fertilization/feeding (cereals)	550	2000
4	Fertilization/feeding (pellets)	2400	

"The daily fish production decreases when the biomass approaches the limit of the carrying capacity of the pond. It is advised to remove the fish from the pond and start again with small fish. In this way it is possible to have 2-3 growing seasons each year."

Van der Lingen believes that the important thing is the weight of the fish stocked, not the number.

STOCKING RATE: HOW MANY FISH TO STOCK?

The stocking rate of fish is closely bound with pond productivity. Rate of stocking of fish is extremely important at all levels of fish culture.

Stocking (Rate)	Result
Too few stocked	Large size fish but low production
Too many stocked	High production but very small fish
Optimum no. stocked	High production of marketable size fish.

How to determine optimum rate of stocking?

To determine optimum rates of stocking for various levels of management, a range is stocking rate must be tested to enable determination of the maximum rate producing desirable size fish in a minimum period of time.

With supplemental feeding, large number of fish can be stocked.

How to calculate the stocking rate?

Stocking rate (no.) =

Growth target or total productivity (in kg) + expected individual growth target (in kg) mortality (%).

The most economic stocking rate is not necessarily that results in the highest average growth rate but rather that which results in the highest yield per unit area.

As for example:

Table 8

No. of Stocks / ha	*Ave. growth rate / day*	*Total yield / day*
2,000	6 g	12 kg
3,000	5 g	15 kg
4,000	3 g	12 kg

Stocking rate also depends on the market size of the fish. The smaller the market size (fish), the higher the stocking density.

GROWTH RATE OF FISH

Fish is the final result of a complex biological cycle in a pond. Fish growth depends on many factors:

(A) Factors of internal origin:

(1) Genetic character (heredity)

(2) Physiological state (health)

a. Ability to utilize food

b. Resistance to disease

(B) Factors of external origin:

(1) Quality of water

(2) Optimum water temperature.

(3) Available oxygen

(4) Available food (quality and quantity)

(5) Metabolite level

(6) Space factor (living space)

The principal rules which determine the growth of fish are:

A. The maintenance need depends on the activity or relative size of fish.

B. Growth rate is relatively greater in small size fish. Age is of secondary importance e.g. rapid growth of stunted fish.

C. Under normal conditions the growth requirements in relation to maintenance requirement lies between 1/1.5 and 1/3.2.

The larger the fish, the higher is their absolute potential growth capacity. The absolute potential growth rate however, does not increase in a direct proportion to the increase in weight of the fish, but a rather slower rate. This means that the relative growth rate i.e. the growth per unit weight decreases with the increase in the weight of the fish.

Table 9

Example		% growth /body weight	weightgain / day
g i h grow	g/day	g/day	g/ g
g i h grow	g/day	g/day	g/ g
g i h grow	g/day	g/ day	g/ g

WHEN TO HARVEST FISH?

For every species it is necessary to find out the fast growing period when the food conversion ratio is most efficient. The crop should be harvested when FCR starts getting high.

Growing very large size fish is uneconomical. Large size fish eat more and grow less. So an intelligent farmer should harvest fish as soon as it reaches minimum, marketable size. The stock thus removed should be replenished. Higher production is obtained by continuous harvesting and replenishment of stock.

Harvest relatively smaller fish. Don't wait long for the stock to grow big.

Raising fry is more profitable than raising fingerlings. Also, raising fingerlings is more profitable than raising marketable-size fish.

What is the principle behind it?

Large number of small fish can better utilize available food in a pond than small number of large fish.

CONTROL OF PREDATORS AND COMPETITORS IN PONDS

High production in fish depends to a great extent on the management skill on the fish farmer.

Predators are those, which prey upon the cultivated species resulting in heavy mortality (low survival) and poor production.

Important predators:

Fish: Piscivorous fishes e.g. sea bass, dalag, tarpon, ten pounder etc. (Small fishes, minnows and their fry are predators of post larvae in nurseries)

Amphibian: Frogs, and frog larvae

Reptiles: Water snakes, turtles

Birds: Kingfisher, heron, cormorant,

Mammal: Otter

Crustacea: Crabs

Insects: Water bugs, larvae of diving aquatic beetles, nymphs of dragonfly.

Competitors

Weed fishes (minnows and misc. small fishes) e.g. barbells, clupeids, tilapia, therapon, gobies. They compete for natural food as well as for feeds. Also they compete for pond space and available oxygen. Also accumulation of (carbon dioxide and nitrogenous) metabolites due to presence of large no. of weed fishes.

Methods of control

A. *Prevention:*

1. Dewatering and complete drying of pond bottom
2. Proper maintenance of embankments and gates
3. Effective screening of inlets and outlets.

B. *Physical control:*

1. Trapping
2. Use of bird scaring device
3. Wire net fencing
4. Cross-wire above the pond/nets over nurseries/small ponds
5. Watchman: Against poaching.

C. *Chemical control:*

1. Poisoning: Use of fish poisons of plant origin
   Use of Chemicals (selective).

POOR MANAGEMENT: LOW PRODUCTION

Low survival/slow growth: due to : predators / competitors

How to Eradicate Unwanted Fishes?

- By repeated through netting
- Draining ponds/ drying pond bottom
- Preventing entry of unwanted fishes
- By use of fish toxicants.

1. Fish poison of vegetable origin

Teaseed cake (Camellia sinensis): 75-100 ppm: toxicity 10-12 days (A.1) Saponin - 10 ppm kills fish: 11ppm shrimp survives.

Tobacco dust *(Nicotiana tabacum)*: 200-400 kg/ha m Nicotin (a.i).

Mohua oil cake *(Bassia latifolia)*: 4-6% saponin (a.i):

200-205 ppm complete kills 6-8 hr. Toxicity 15-20 day. Also fertilizer

Derris root power *(Derris robusta):* Rotenone (a.i) 5 % in DRP, DRP 2-5 ppm (Rotenone 0.1-0.25 ppm): more effective at higher temperature: Toxicity 1-2 weeks counte-acted by $KMnO_4$ (2ppm against 0.05 rotenone).

Seed Root-bark and sten-bark of *Barringtonia acutangula* 10-20 ppm kills fish within 1-2 hr. plankters not affected toxicity: 48 hr.

Speed power of *Croton tiglium:* 3-5 ppm and *Millettia piscidia* 4-5 ppm and unripe fruit power of *Randia dumetorum* 15 ppm; seed husk of *Tamarindus indica*: 5-10 ppm.

2. Chemicals (Pesticides not recommended)

Anhydrous ammonia: fish kills at 20-25 ppm toxicity 5-6 week fertilizer and also a herbicide.

Ammonium sulfate: 50-100 kg/ha with lime, which serves as a catalyst.

Bleaching power (Calcium hypochlorite): 25-30 ppm in 3-4 hr.: Toxicity: 7-8 days.

AQUATIC INSECT CONTROL

Soap oil emulsion: 54kg/ha cheap vegetable oil mixed with 18kg/ha cheap soap sprayed over the water surface.

Kerosene oil: 80-1001/ha kills insects, also zooplankton .

LSD (Light speed diesel oil): LSD (11) + emulsifier hyoxid 1011 (0.75ml) + 40 ml water @ 52 litre/ha.

POND SOIL

Soil is very important for fish production in ponds due to its ability to store and release nutrients necessary for growth of plankton, the natural food of the fish. It is the main source of dike-building materials of ponds. Also it regulated water quality.

TYPES OF SOILS: (Extreme type)

1. *Acid Soils:* Acidic soils having very low pH (acid sulfate soil)
2. *Saline soils:* Saline soils are those which have excess of soluble salts
3. *Alkaline soils:* They are rich in basic salts (usually in low rainfall areas).

Physical Properties of Soil

A. *Soil texture:* Relative proportion of sand, silt and clay in a soil mass is called soil texture.

B. *Soil structure:* The arrangement of the individual soil particles with respect to each other into a pattern is called soil structure.

Soil texture: The following three textural classes are important:

A. *Fine texture:* Clay, silty clay, clayloam, silty clay loam and sandy clay.

B. *Medium texture:* Silt, silt loam, and loam

C. *Coarse texture:* Sandy loam, loamy sand and sand

The sand, silt and clay are differentiated from one another on the basis of their particle size and feel:

(1) Sand: 0.02-0.2 mm. Gritty when rubbed, non-plastic and non-sticky.

(2) Silt: 0.002 - 0.02 mm. Smooth, powdery when rubbed; slightly plastic and sticky when wet.

(3) Clay: Less than 0.002 mm. smooth, sticky, plastic when wet. Forms very hard clod when dry and remains suspended in water for long period of time.

Chemical Properties of Soil

Macro and micro-nutrients present in soil are essential for fish production. Macro-nutrients are N, P, K, Ca, and Mg. Micro-nutrients (trace elements) are Fe, Mn, Cu, Zn, Mb, Mo, Co etc.

LIMING

Liming in fishponds is advocated for several reasons, which are beneficial to fish culture.

- It is used mainly to correct acidity of pond soil and water. It raises the low pH of acid waters to slightly alkaline and thus increases the productivity of water.
- It increases the alkaline reserve of mud and water and by its buffering action prevents extreme changes in pH.
- It creates a carbon dioxide reserve and ensures adequate calcium for plant and animal nutrition. Pond containing lime is likely to be more fertile than one without.
- It counteracts harmful effects of excess of Mg, Na and K ions and those of humus acid and sulphuric acid.
- Most organisms responsible for nitrification of NH_4^+-compounds require large amount of active Ca.
- It assists in the release of nutrients from the pond soil and promotes bacterial breakdown of waste materials.
- It improves structure of fine textured soils by increasing the organic matter content and by flocculation of Ca-saturated colloids.
- It acts as a powerful disinfectant and kills by caustic action harmful organisms at the pond bottom.

Caution

Lime is toxic and so ponds should not be stocked with fish until at least a fortnight after lime treatment. Excess of lime may bind available PO_4^3 into insoluble Ca_3 $(PO_4)_2$

Types of lime to be used

Calcium oxide (CaO): also known as quick lime, unslaked lime or burnt lime. It contains 71.4% Ca.

Calcium hydroxide [$Ca(OH)_2$]: also called slaked lime, hydrated lime or builder's lime. It contains 54.09% Ca. It is prepared by hydrating CaO.

Calcium carbonate [$CaCO_3$]: available as crystalline $CaCO_3$ in nature and is termed calcite or calcitic limestone.

NITROGEN CYCLE IN AQUATIC ECOSYSTEM

The primary source of inorganic nitrogen is by fixation of molecular N_2 from atmosphere by biological, meteorological and industrial processes.

The transformation of N_2 and other nitrogenous compounds in water could be outlined as follows:

1. **Nitrogen fixation:** (Process of directly utilizing atmospheric N_2). N_2 is fixed in waters mainly by the heterocyst containing blue green filamentous algae such as Anabaena, Nostoc etc. Anabaena associated with floating weed Azolla also fixes N_2. When these organisms die, they contribute mineralized nitrogen. The common heterotrophic bacteria Azotobacter and Clostridium and photosynthetic bacteria also fix N_2.
2. **Aminization:** This process involves breaking down of protein from organic matter (dead materials) into amino-groups.
3. **Ammonification:** This process involves deamination of proteins and other nitrogenous compounds by microbial activity and production of NH_3-N. NH_3 is released to the environment.

   NH_3 + H+ NH_4 +
4. **Nitrification:** This is the process of converting ammonia into nitrites and nitrates. The nitrifying bacteria nitrosomonas oxidizes NH_4^+ to NO^2 - and obtain energy:

$NH_4^+ + 1\ 1/2O_2$ $2H^+ + NO_2^- + H_2O$ (1)

Oxidation of NO_2 Proceeds further to NO_3^- by the bacteria nitrobacter

$NO_2 + 1/2O_2$ NO_3^- (2)

5. **Denitrification:** This is a heterotrophic process which results into the liberation of gaseous forms of nitrogen as metabolites and is lost from the system. Some Facultative anaerobic bacteria e.g. Pseudomonas and Achromobacter are able to denitrify nitrate to nitrogen gas.

NH_3 (Gas)

NO_3^- NO_2^- N_2 (Gas)

N_2O (Gas)

6. **Some Vital microorganisms (Functional Organisms)**

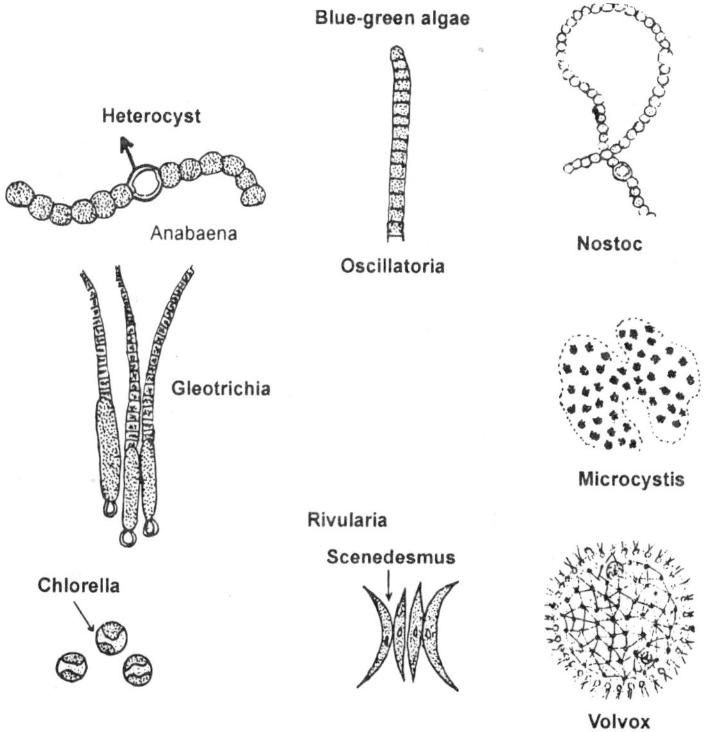

Fig. 1

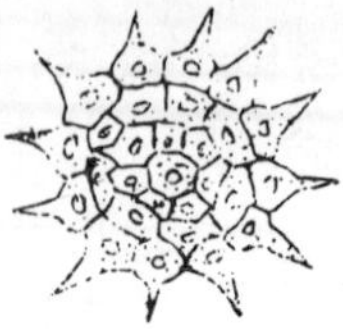

Pediastrum

Green algae : Desmids

Cosmarium Closterium

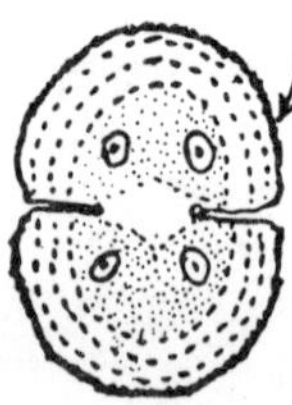

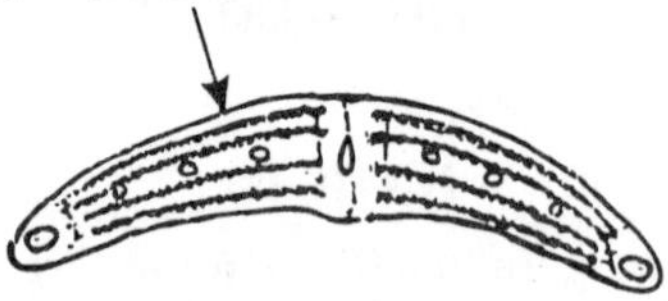

Euastrum

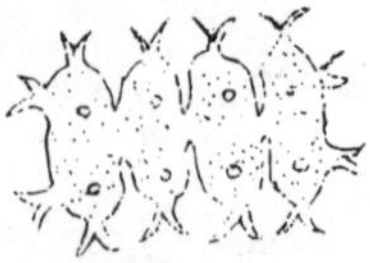

Xanthidium

Green algae : Filamentous forms

Ulothrix

Zygnema

Spirogyra

Fig. 2

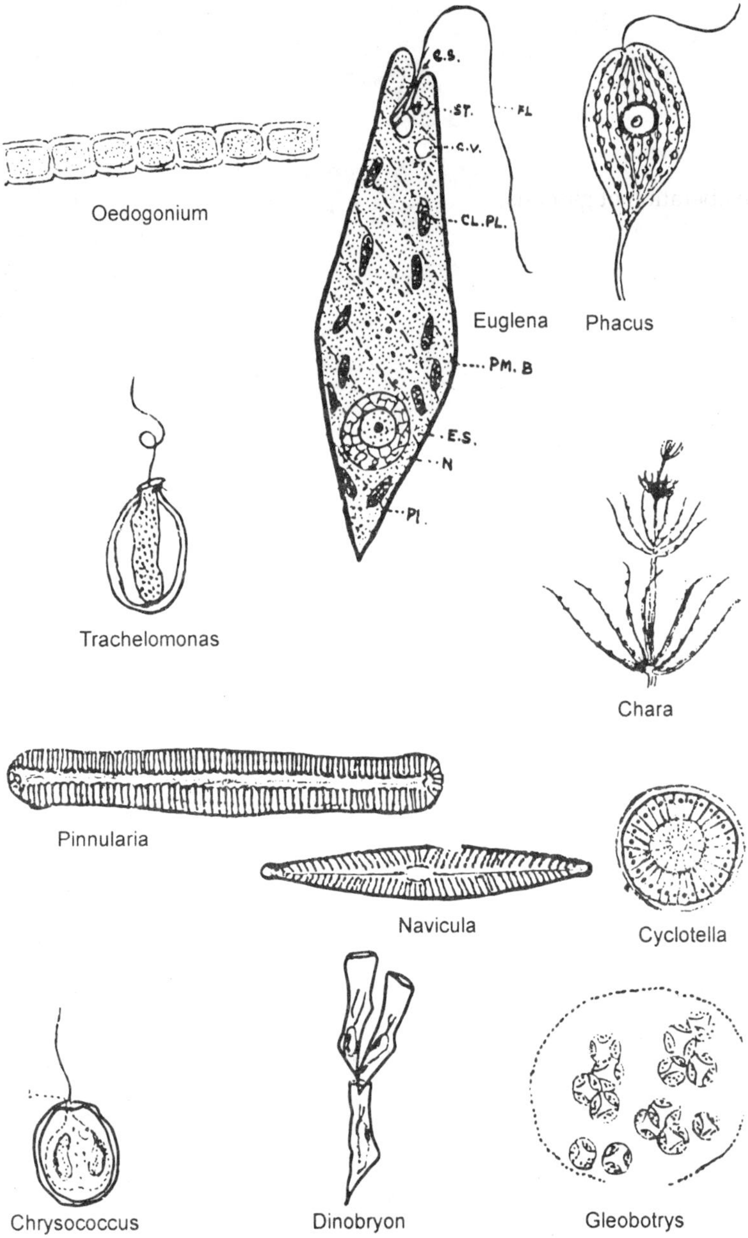
Oedogonium
e.s.
ST.
FL
C.V.
CL.PL.
PM.B
E.S.
N
Pl.
Euglena
Phacus
Trachelomonas
Chara
Pinnularia
Navicula
Cyclotella
Chrysococcus
Dinobryon
Gleobotrys

Fig. 3

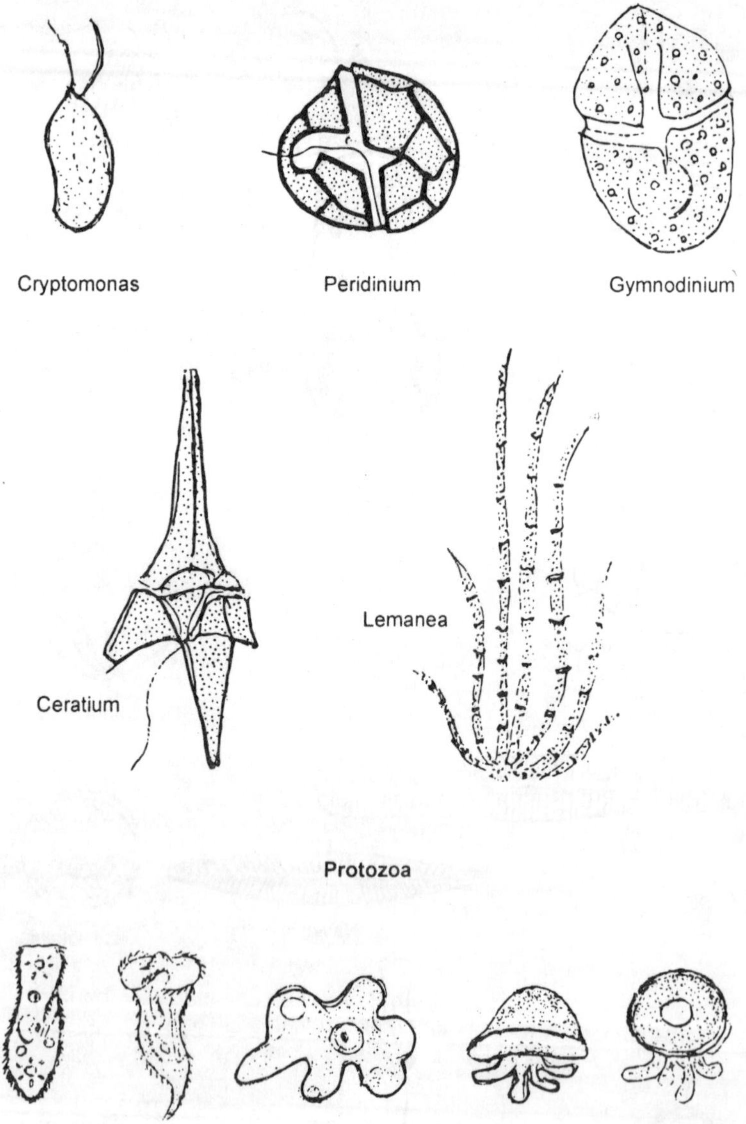
Cryptomonas
Peridinium
Gymnodinium
Lemanea
Ceratium
Protozoa
Paramecium
Stentor
Amoeba
Arcella

Fig. 4

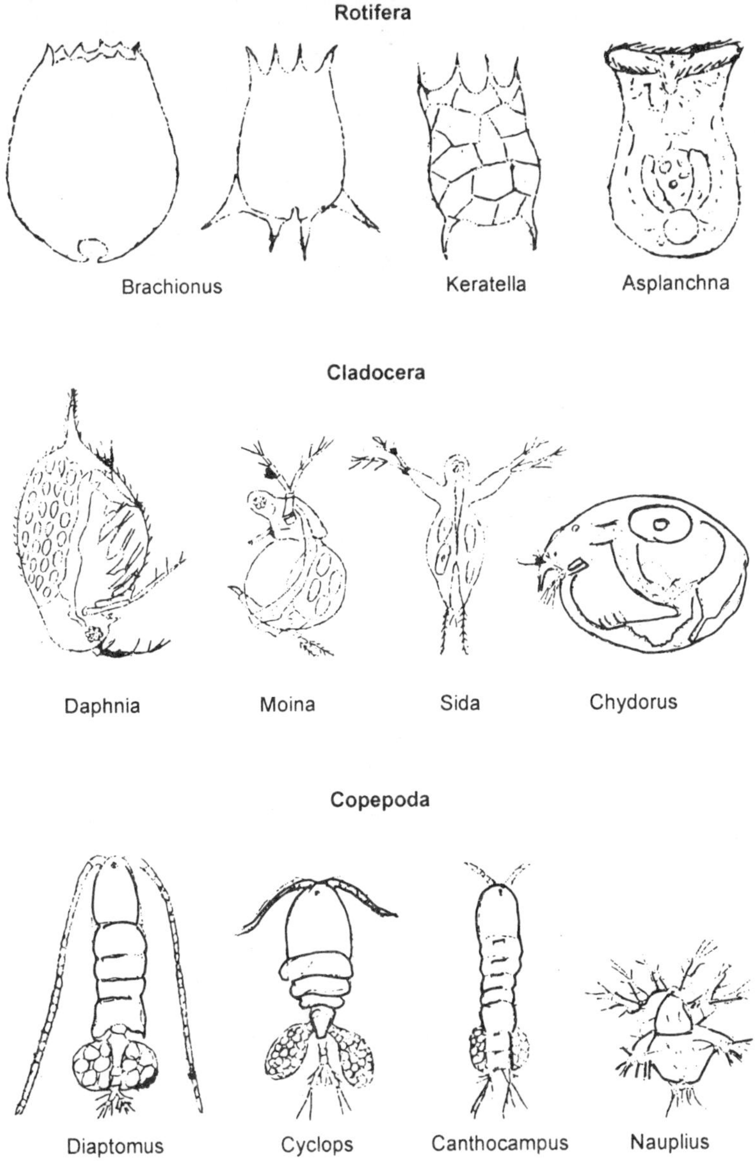
Rotifera
Brachionus
Keratella
Asplanchna
Cladocera
Daphnia
Moina
Sida
Chydorus
Copepoda
Diaptomus
Cyclops
Canthocampus
Nauplius

Fig. 5

11
REPRODUCTION IN FISHES

Reproductive Organs (Gonads)

Testes in males	Male sex cell: Sperm
Ovaries in females	Female sex cell: Ovum (egg)

Main Types of Reproduction

(A) *Bisexual:* Sperms and ova develop in separate males and females; very common

(B) *Hermaphroditic:* Both sexes occur in one individual; but rarely self-fertilization takes place. Usually sperms and ova do not develop simultaneously.

(i) *Protandrous:* Fishes which are males fist e.g. some basses

(ii) *Protogynous:* Fishes which are females first e.g. Grouper (Lapu-lapu)

Reproductive Patterns in Fishes

A. *Oviparous:* Fishes that lay their eggs outside the body for development; fertilization takes place outside in water e.g. carps and majority of bony fishes.

B. *Ovo-viviparous:* Young ones in these fishes are born alive. Fertilized eggs develop inside the body (oviduct), but the embryo derives its soul nourishment from the egg and not from the mother, e.g. mosquito-fish, guppy

C. *Viviparous:* The young are born alive. Fertilized eggs develop in the uterus and derive nourishment from the mother, e.g. some sharks.

Sexual Reproduction

It requires two parents. Each parent contributes a sex-cell (gametes). Haploid gamete × haploid gamete combine to form diploid zygote a new individual develops.

Repression Factor

Mature fishes when kept overcrowded in a body of water. Secrete or excrete certain substances (hormone like in nature) in the water which inhibit reproduction of the species.

This repressive factor may be specific or may effect other species. By the use of this information it is possible to prevent or retard or to obtain reproduction in many species of fishes.

Gynogenesis

It is a type of reproduction in which insemination takes place but the sperm uncleus does not take part in the development and the ovum nucleus produces a female individual with diploid number of chromosomes.

It occurs naturally in a crucian carp (Carassius auratus gibelio: Silver goldfish) and also in *Poecilia formosa*.

Parthenogenesis

Unisexual reproduction involving the production of young by females not fertilized by males. A parthenogenetic egg may be diploid or haploid. e.g. common in rotifers, cladocerans, bees, etc.

Endocrine System

It regulates and integrates some body processes by the production and secretion of hormones.

Endocrine gland: Pituitary, thyroid, para-thyroid, adrenal, pancreas, ovaries and testes. These are ductless glands which synthesize hormones, and discharge these directly into the blood.

Hormone: Hormones are chemical substances produced by one part of the body (endocrine glands) and transported to another (target organ) where they exert their effects. They act as messengers.

Pheromone: It is a substance secreted by a fish/animal which when excreted, provokes specific reactions in other individuals ordinarily of the same species. It appears to be a sex-steroid.

Steroid: Steroids are fat soluble compounds. Hormones produced by gonads, such as, estrogen, progesterone and testosterone are steroids.

Hypophysis (pituitary gland): Embryo-logically formed by depression of the neural ectoderm (nervous part) and by thickening of the ectodermal lining of the roof of the mouth oral ectoderm (glandular part). The secretions of this endocrine master gland regulate a wide variety of physiological processes in fishes.

Vitellogenesis: Formation (deposition) of yolk granules/ vesicles in the egg (ovum)

Follicle: It consists of group of epithelial cells surrounding a developing egg.

Corpus luteum: Yellow body formed in the ovary after the follicle ruptures and ovum released from the ovarian wall. It secretes progesterone.

The complexity of endocrine-logical inter-relationships makes it difficult to state unequivocally the exact function of different hormones.

12
METABOLIC PRODUCTS OF FISH AND THEIR EFFECTS OF THE GROWTH OF FISH

Metabolism

It is a physiological process and is the result of all the chemical and energy transformations that occur within the living organisms. All phases of metabolism require energy, which animals obtain from food. Energy is expended for the maintenance and for growth. It has two processes :

a. **Anabolism:** The assimilation and storage of energy in protein, fat and carbohydrate.

b. **Catabolism:** The transformation of energy stored in protein, fat or carbohydrate forms into energy. Or in other words "The release of body energy, the building and repairing of body parts and the excretion of waste materials all put together is Metabolism".

Products of metabolism must be eliminated from the body because they are or may become toxic to the animals producing team. In fish, the metabolic products are excreted mainly through the gills. These are carbon dioxide, ammonia, urea, amine and amine-oxide derivatives. Remaining products such as, creatine, createnine and uric acid are eliminated through the kidneys. These wastes are excreted directly into the water in which the fish lives. Accumulation of metabolites in culture water inhibit growth of fish and very often become toxic to fish.

Metabolites: The nitrogenous metabolites released by fish and other aquatic organisms are: (a) Ammonia (NH_3), (b) Creatine, (c) Creatinine, (d) Urea, (e) Uric acid and Amino acids. In addition, nitrogenous compounds are also released during the bacteriological decomposition of plants and animals. These are next transformed into ammonia which undergoes nitrification of nitrate (NO_3) through a nitrite (NO_2) intermediate (see Nitrogen Cycle). CO_2 is a non-nitrogenous metabolite.

Ammonia: NH_3 is one of the chief products of protein metabolism. Fish is very sensitive to ammonia and ammonium compounds. For fish culture 0.1 ppm of NH_3 is the maximum to be tolerated ordinarily. Above this level NH_3 appears to be toxic to many fishes and 1 ppm is definitely an indication of pollution. As NH_3 concentration rises in water, the O_2 content decreases and CO_2 content of blood increases. The fish loses its ability to utilize O_2 for respiration and when the concentration reaches 1 ppm, the hemoglobin of blood loses its ability to unite with O_2 or to liberate CO_2 and fish dies of asphyxia. An increase in water temperature and pH also aggravates this effect. Thus, we find that NH_3 and to some extent CO_2 when accumulate, upset the normal metabolism of fish. When metabolism is affected, growth rate is also affected. CO_2 can easily be eliminated through aeration or utilized through photosynthesis. But nitrogenous waste products cannot be eliminated through aeration alone. Dilution of water however, helps dilution of metabolites. The rate of change of water has to be increased to ensure greater dilution of metabolic products and ultimately their elimination.

Un-ionized ammonia is toxic to fish but ionized NH_4 is not toxic. NH_4^+ plus un-ionized NH_3 determine the total ammonia-nitrogen. In channel catfish, 0.12mg/l of NH_3-N causes reduced growth and gill damage, whereas a concentration of 0.52mg/l of NH_3-N causes 50% reduction of growth and no growth at 0.97 mg/l. In running water fish culture these growth-inhibiting metabolites are constantly washed out and cannot affect the growth rate. The rate of flow is important as there is a positive correlation between yield of fish and rate of flow of water. In Japan, the yield of carps in running water was found to be 100-1000 times higher than in standing water where metabolites accumulated.

The growth inhibiting character of fish metabolites is utilized in stunting fish. Keeping fish stunted is a traditional practice in India and a few other countries for making fish seed available throughout the year, especially for the species who are seasonal breeders. Fish seed is heavily stocked in ponds and are fed adequately. These fish remain healthy but do not grow because of the presence of metabolites in water. Thus fish seed could be made available whenever required.

13

Cage Culture in Running Water

Advantages

- Productivity 10-20 times higher than ponds for comparable inputs and area.
- Eliminates inherent drawbacks of pond culture such as, loss of stock due to flooding, seepage and evaporation. There is no need for water replacement.
- No dependence on unfavourable pond soil characteristics.
- No problem of pond excavation in marshy/water-logged areas; high cost of pond construction.
- Cages can be located conveniently near urban areas ensuring easy marketing.
- Increasing demand on lands making suitable land unavailable for ponds.
- Flowing water gives higher production by eliminating nitrogenous waste products.

Criteria for Achieving High Production through Running Water Culture

- Abundant supply of suitable water
- Optimum rate of flow
- Oxygen-rich water
- Effective elimination of metabolites
- High protein- rich compete diet
- Quick growing cultivated species
- Efficient FCR
- Availability of seed
- Optimum water temperature
- Intensive stocking
- Feeding *ad libitum*.

14

SIMULTANEOUS CULTURE OF RICE AND FISH

- Fishes control weeds from rice fields
- Fishes control insect pests
- Biological control of mosquito larvae by larvivorous fish
- Some fishes control snails harmful to rice plants.
- Paddy tillering caused by grazing of fish
- Fish excreta fertilize paddy field
- Fish is a source of animal protein
- Grain output is increased in presence of fish
- Highly fertile paddy bottom: Good growth of fish-food organisms
- Concurrent culture: most economical utilization of land
- To raise double crop, farmer pays more attention: Better result

Disadvantages of Growing Double Crop

- All rice fields not suitable: needs adequate water depth
- Needs high dykes, trenches: additional cost
- Needs adequate drainage and extra water: additional cost
- Fish should tolerate high temperature, low oxygen content and high turbidity
- Low survival of fish due to predation/pesticide
- Fish restricts use of insecticide/weedicide/chemical fertilizer.

For concurrent rice fish culture, it is essential to have: Strong water-tight embankments, trenches and canals and inlets and outlets provided with screens.

These physical modifications depend on:

- Water supply
- Rainfall and
- Climatic condition.

Rotational Cropping System

Rotation of Rice and Fish

(rice-rice-fish or fish-fish-rice or fish-rice-fish)
(Tables 6 and 7)

Advantages

- Allows better cultural techniques for both
- Pesticides can be used for higher yield of rice
- Permits greater depth of water for better growth and survival of fish
- Trenches and canals not needed: Saves space and money
- Better control of insect pests: Life cycles are disrupted
- Residues from fish culture act as fertilizer for rice plants
- Rice stubbles grow fish food organisms by rotting after being submerged

Constraints

- Water requirement very high
- Only selected rice varieties can be used
- Possible in areas where chance of flooding is minimal
- Greater water depth required

Table 10 : Some Figures of Fish Yields in Rice-Fish Culture
(Mostly production through traditional culture)

Country	*Species*	*Yield (kg/ha/yr)*
Japan	(Common carp) *Cyprinus carpio*	100-200 (kg/ha/yr) 700-1,800 (kg/ha/yr) (W/feeding)
Indonesia	*C. carpio*	75-100 (kg/ha/yr) 352.8 (kg/ha/yr) (3-4 mos)

Contd...

Table 10 – *contd...*

Country	*Species*	*Yield (kg/ha/yr)*
China	Grass carp & other Chinese carps	100-150 (kg/ha/yr) 937.5 (kg/ha/yr)
India	Seabass, mullets, catfish, shrimps	100-200 (kg/ha/yr)
Malaysia	*Trichogaster pectoralis* (Sepat siam)	10-400 (kg/ha/yr) (Coche, 1967)
Thailand	Common carp	60-120 (kg/ha/yr) 3-4 mos. 210-250 (kg/ha/yr) 6 mos. (with feeding)
Philippines	*Java tilapia* *Nile tilapia* *Nile tilapia & C.carp*	94-200 (kg/ha/yr) 3-4 m 100 (kg/ha/yr) 3-4 m 187 (kg/ha/yr) 3-4 m 293 (kg/ha/yr) 3-4 m (Cruz, 1984)
Taiwan	*C. Carpio*	235 (kg/ha/yr) (Chen, 1953)
Vietnam	*C. Carpio*	50-130 (kg/ha/yr) 10m
U.S.A	Big-mouth buffalo Channel catfish Common carp	119 (kg/ha/yr) (Average) (Coche, 1967)
Madagascar	*Java tilapia* Common carp Goldfish	200-205 (kg/ha/yr) (Coche, 1967)

Table 11 : Toxicity of Commonly Used Insecticies

Sl. No.	*Insecticide (common name)*	*Formulation*	*96 h LC_{50} PPM*
A.	**Test Fish:** *Clarias batrachus*		
1	Diazinon	Emulsifiable concentrate (EC)	10.0
2	Chlorpyrifos	EC	1.0
3	Fenitrothion	EC	2.0
4	Fenthion	EC	1.8
5	Malathion	EC	1.3
6	Carbofuran	Granular	3.2
7	Carbaryl	Wettable powder (WP)	20.0
8	X-BHC	EC	0.13
9	Endrin	EC	0.00063
10	Emdosulfar (Thiodan)	EC	0.0028

Contd...

Table 11 – *contd...*

Sl. No.	*Insecticide (common name)*	*Formulation*	*96 h LC_{50} PPM*
B.	**Test Fish:** *Cyprinus carpio*		
11	Dichlorvos	EC	2.30
12	Fenitrothion	EC	3.40
13	Methomyl	WP	5.80
14	Carbaryl	Soluble Powder	8.20
15	Endrin	EC	0.0040
16	Endosulfan	EC	0.0092

Organophosphates (1-5 and 11-12)
Carbonates (6-7 and 13-14)
Organochlorines (8-10 and 15-16)

Table 12 : Species of Fish Used in Rice-Fish Culture

Scientific name	*Common name*
Oreochromis mossambicus	Java tilapia
Orechromis niloticus	Nile tilapia
Cyprinus carpio	Common carp
Catla catla	Catla
Labeo rohita	Rohu
Cirrhinus mrigala	Mrigal
Ctenopharyngodon idella	Grass carp
Puntius gonionotus	Tawes
Carassius auratus	Goldfish
Osteocheilus hasselti	Nilem
Helostoma temmincki	Kissing goramy
Trichogaster pectoralis	Sepat siam
Osphronemus goramy	Giant goramy
Chanos chanos	Milkfish
Mugil parsia	Parse
Liza tade	Grey mullet
Mystus gulio	Catfish
Clarias batrachus	Air-Breathing catfish
Channa striatus	Mudfish
Anabas testudineus	Climbing perch
Lates calcarifer	Seabass
Ictalurus punctatus	Channel catfish
Ictiobus cyprinellus	Bigmout buffalo

15

Aquaculture Detailed in Transparencies

Increase Production – Biotechnology – Productivity – Seaweed Culture – Human Population and Environmental Exploitation

Aquaculture Manual : The Transparencies Used by Prof. Chaudhuri as Visiting Professor in the University of the Philippines (1988-1993)

AQUACULTURE FOR INCREASED PRODUCTION OF FISH

How to increase production ?

Four I's

- Increasing the Area for Culture
- Intensification of Culture
- Improvement of Technology
- Introduction of Non-Conventional Methods.

Intensity of Culture

- Extensive (improved traditional)
- Semi-intensive
- Intensive
- Fish - a good source of high quality animal protein. Remarkable progress has been made in (a) fish production, (b) Food fish production.

THREE-DIMENSIONAL AQUACULTURE

In this type of aquaculture the entire body of water from top to bottom is utilized for culture. The commodities are: (a) Molluscs (b) Seaweeds

A. Molluscs

Cultured species are mostly the bivalves which are filter feeders such as, Oysters (e.g. *Crassostrea gigas*), mussels (e.g. *Perna viridis*), clams, cockles and scallops.

These are cultured in coastal shallow waters (mostly in the intertidal zone) directly on the seabed or using stakes or poles, racks and trays. In deeper waters, floating rafts or long lines are used (culture by suspension).

Types of molluscs culture

1. Bottom culture
2. Pole culture
3. Tray culture
4. Raft culture
5. Long line culture

B. Seaweeds

Seaweeds, such as, *Eucheuma gracilaria*, Porphyra and others are cultured in protected bays is shallow coastal waters.

1. Bottom Culture: *Eucheuma*, *Gracilaria* and *Caulerpa* are cultured in bottom habitats. The net or long line methods are used.

2. Culture by suspension: The red seaweeds *Porphyra* are cultured on rectangular (horizontally placed) nets tied to stakes. Long line method is used for Undaria and Laminaria.

Non-conventional: Vertically integrated fish farming systems.

Three dimensional aquaculture: Molluscs/seaweeds

Monoculture	and	polyculture of species.
↓		↓
Modular system		Multigrade system
↓		↓
Multi-size stocking		Plyculture of mixed size & mixed age

Mono-sex culture: Culture of forage-cum-predatory species.

Int .grated farming systems

Recyclying of wastes: Aeration: Elimination of metabolites

Running water fish culture: very high production

Cage and Pen culture

Recirculated culture system

Sophisticated aquaculture;

Super intensive culture of shrimps (dollar earner)

- *Penaeus monodon* (Tiger shrimp)
- *Penaeus japonicus* (Kuruma shrimp)
- *Penaeus indicus* (White shrimp)
- *Penaeus merguiensis* (Banana shrimp)

Eye-stalk ablation technique for maturation hatchery and grow-out ponds /tanks.

Improvement of stock by genetic selection and hybridization

BIOTECHNOLOGY

Def. 1 "Controlled use of biological agents, such as, micro-organisms or cellular components, for beneficial use."

Def. 2 "The collection of industrial processes that involve the use of genetically engineered micro-organisms".

Def. 3 "The application of scientific and engineering principles to the processing of materials by biological agents to produce goods and services".

[Source: Committee on Bistech, U.K.]

Biotechnology : Offers great promise to solve biological problems in terms of food production. These approaches have largely been applied to agriculture and animal husbandry, but very little progress so far has been made in enhancing fish production.

POTENTIAL FOR USING BIOTECHNOLOGY

Approaches in aquaculture species

(Three main approaches)

- Improved diagnostic reagents:
- Hormones, feeds and vaccines:
- Genetic manipulations.

Hormonal sex control in aquaculture

(To prevent overcrowding in ponds due to prolific breeding by fishes like tilapias).

Sex Reversal

(Production of single sex stocks by administration of hormones (steroids) to Juvenile fishes).

In *Oreochromis niloticus* males are produced by feeding young fry with 17°C Methyl testosterone. Males grow bigger and faster than females in Tilapia. In fishes where the female is bigger and grows faster the sex-reversed females are produced by feeding early fry with estrogen 17B- estradiol benzoate or diethyl stibestrol.

Sterilization

Effected by administration of higher dose of Methyl Testosterone.

Biotechnology in single-celled algae culture

Principles of biotechnology can be readily applicable to single-celled algae which are used as feed in aquaculture. The green alga *Chlorella* and the cyano bacterium *Spirulina* can also be used as a valuable human food source.

- These algae could be easily manipulated genetically.
- Hybrid strains produced by protoplast fusion technique.

Cryopreservation Technology

Cryopreservation means preservation of gametes and embryo at below freezing temperature for ready availability of genetic material for breeding and research purposes.

While cryopreservation of sperms (Milt) of fishes has been very successful, the same for eggs and embryos of aquaculture species has met with very little success.

Application of Biotechnology to Produce Improved Feeds:

Fish production can be increased by development of improved feeds such as:

- Development of new artificial diets, especially for the filter feeders (oysters and mussels) through micro-encapsulation of dietary compounds.

- Development of food additives such as, anabolic steroids and thyroid hormones, have led to enhanced growth of salmonid fishes.
- Improved culture of single celled feed organisms, such as algae, yeast and bacteria also helps increase in aquaculture production.

USE OF BIOTECHNOLOGY

Stock Assessment in Marine Environment

Molecular technique, such as, DNA finger-printing, is an useful technique of stock assessment of migratory fishes and also for species identification in the sea. For comparison of mitochondrial DNA from different individuals, a method has recently been developed in mapping the mitochondrial DNA genome. Analysis of mitochondrial. DNA offers a valuable mechanism in assessing the population structure in marine fishes and invertebrates.

Manipulation of Chromosome Sets

By Chromosomal Manipulations

Polyploidy, Gynogenesis and Androgenesis could be induced in fish for enhancing fish production.

Polyploidy

Production of individuals having extra sets of chromosomes by subjecting fertilized fish eggs to heat or cold shocks. They are sterile forms and this advantage is utilized in aquaculture in getting higher production.

Gynogenesis

It is induced in fishes by stimulating parthenogenetic development of fish eggs by artificially inactivated spermatozoa. Sperms are irradiated by X-rays or UV-rays and their genetic contents (DNA) are destroyed. Fertilization occurs without any genetic contribution from the sperm (all maternal inheritance)

Androgenesis

In androgenesis, fish ovum is irradiated with X-ray of UV-ray and its genome is destroyed. Fertilization occurs without any genetic contribution occurs without any genetic contribution from the ovum and the offspring is a male (all paternal inheritance).

Advantages of Andro- or Cyto-genesis

(a) Useful in the generation of inbred lines, progeny testing and selective breeding.

(b) Production of Uri parent or mono-sex culture in nature, gynogenesis occurs in crucian carps.

Parthenogenesis

Unisexual reproduction involving production of young by females without fertilization by males. A parthenogetic egg may be diploid or haploid.

Trangenesis (gene transfer)

Transgenesis is the process of *in vitro* transfer of exogenous and usually heterogenous genome of a fish or any organism.

Such a fish is called a transgenic fish.

Transgenic fish

Transgenetic studies offer great promise to produce *Super fish* and have high potential when applied for the benefit of aquaculture.

However, some technical problems need be solved before gene transfer is widely accepted in aquaculture.

Genetic Manipulations

Genetic manipulations mean changes in the genetic make up of any organism by application of improved techniques. Fishes can tolerate a wider range of genetic manipulations than terrestrial farm animals. Fishes have an additional advantage that they produce large number of gametes and because of *in vitro* fertilization of eggs, genetic manipulations are easier.

Various methods used are:-

- Manipulation of chromosome sets
- Transgenesis or gene transfer
- Cell culture and nuclear transplantation
- Cryopreservation technology

Cell Culture Biotechnology

Application to marine macro-algae

- Marine macro-algae or seaweeds could be easily genetically engineered.
- Entire seaweed plant regenerated from isolated cells and protoplasts.
- Development of new strains of seaweeds through hybridization of gametes from separate populations.
- Disease resistant strains produced using selection in cell culture.
- Increased production of red seaweed *Porphyra* by use of biotechnology.

Application of advanced technologies in food-fish production

- Induced breeding techniques for seed production
- Selective breeding and hybridization
- Polyculture of compatible spp. for higher yield
- Integrated fish-crop-livestock farming systems
- Running water aquaculture
- Intensification of shrimp culture and improved management techniques for increased production.

BIOLOGICAL PRODUCTIVITY

Biological productivity of any body of water is the end result of the interaction of the organisms present with the surrounding environment. Each organism possesses inherent characteristics and abilities which unable it to exist and reproduce. This is known as the biotic potential of the organism and there is always constant struggle between the biotic potential of the organism. However, conditions in the environment affect and restrict the organism and there is always constant struggle between the biotic potential of the organism with the factors in the environment. Also there is a constant struggle among the organisms themselves (intro- and inter-specific struggle).

Some physical factions of lakes influence biological productivity (BP):

Under strictly comparable condition, the greater the length of the shoreline, the greater the biological productivity of the lake. The increased irregularities of shore line results in:-

1. Greater contacts of water with soil - and more nutrients are released.
2. Increased areas of protected boys and coves. Protected shoal permits:

   a. Growth of rooted plants,

   b. Accumulation of organic bottom materials

   c. Presence of much larger and more diversified fauna.
3. increased shallow areas for growth of rooted vegetation
4. Greater diversification of bottom and margin condition
5. Reduction of the amount of exposed, wore-swept shoal; and
6. Increased opportunity for extensive, close superposition of the photosynthetic zone upon the decomposition zone.

According to their mode of nourishment, the organisms present in a body of water may be classified as:

Autotrophs: [Self-nourishing organism]: Those organisms that are able to prepare organic matter from inorganic with the help of solar energy e.g. chlorophyll-bearing plants.

Heterotrophs: [Exploiters of autotrophs]: Those organisms that are unable to prepare their food and are dependent upon autotrophs for food directly or indirectly, e.g. animals.

THE PARADOX

Despite many significant research breakthroughs, output of fishfarmers is quite low...?

There is a big gap between:

What is achieved by researches and what is produced by fishfarmers.

There is a missing link .... What is that link?

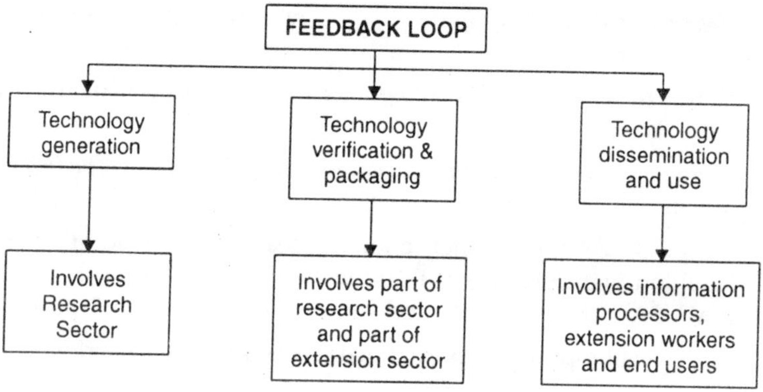

Fig. 6 : The technology transfer continuum

A monitoring mechanism is built into the continuum to provide the forward and backward linkages between the various components.

Technology verification roughly means ... "Testing of research results in farmer's fields". The objectives of the field trials are to:

- Verify or validates technology results.
- Obtain feedback information to guide research programmes.
- Evaluate practices and technologies both technically and economically.
- Better understand traditional farming practices.
- Learn how farmers receive, evaluate and eventually uses the technology.
- Modify technologies for specific areas.
- Introduce new technologies to extension workers and farmers.
- Train professionals working in farmer level experimentation, the research and farmers.
- Serve as an interphase or overlap between research and extension.
- Package the result of verification for dissemination.

SEAWEED CULTURE

Green seaweed: Caulerpa, Ulva, Enteromorpha.... human food.

Brown seaweed: Sargassm ....Source of algin undaria, Laminaria

Red seaweed: *Porphyra, Gracilaria eucheuma, Hypnea, Gelidiella, Gelidium.*

In 1973: 2.4 million metric tons of seaweeds valued at US $ 765 million were produced. Bulk of these - from Japan , China, Korea, USA, Brazil, Norway and some other countries in Europe.

In 1980 Philippines exported over 13000 mt. of seaweed worth 55.6m.

Phyco-colloids from Seaweeds

Agar: Prepared from - *Gracilaria, Gelidium* and *Gelidiella* used for: As culture medium for bacteriological work; manufacture of gelatin, as anti-drying agent in bread and pastries; manufacture of frozen dairy products.

Algin: Prepared from Sargassm.

It is an important component in the manufacture of ice-cream and antibiotics; used in rubber and textile industry.

Carrageenans: Prepared from Eucheuma and Hypnea used in the manufacture of surgical jellies and ointment, cream etc.

HUMAN POPULATION AND ENVIRONMENTAL EXPLOITATION

The balance of nature remains a balance only if there exists a proper give and take relationship among the FOUR GROUPS in Fig. 7.

Of all the living organisms on earth, only man has seriously threatened the delicate natural balance existing among these four groups.

He is now increasing his threats to the point that he threatens not only the existence of non-human organisms but also his own,

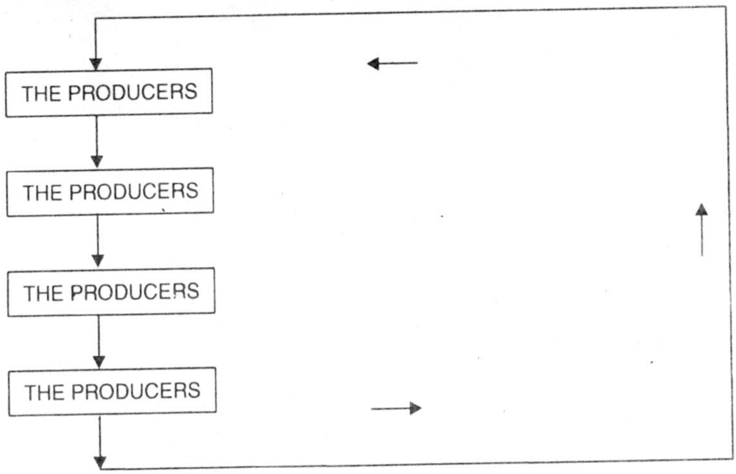

for man is the only organism that pollutes... and also manipulates his environment to serve his own needs and desires.

Lates calcarifer	Seabass
Ictalurus punctatus	Channel catfish
Ictiobus cyprinellus	Bigmout buffalo

Annexure

The Government of India desired that the Professor Hiralal Chaudhuri made a film on *Induced Breeding* to be presented at the FAO World Conference on Aquaculture at Kyoto, Japan in 1976.

Based on the text prepared by the Professor the Information Ministry, Government of India did the filming under the Technical Supervision of Professor Chaudhuri. Except for the photography and music the Professor wrote the dialogue and edited the film. Another film was also produced on "Composite Fish culture under supervision from Institute at Cuttack.

The Professor created lasting interest among the students by displaying the film.

This film was considered the best among 72 films exhibited in the conference.

FILM: INDUCED BREEDING – AN OUTLINE

A common scene: Catching fish in river

Music (m)

A poor catch

(m)

Fish is an important source of protein. The average yield in fish culture is low.

(m)

The flowing water of rivers, where cultivated fish such as Catla, Rohu, and Mrigal breed during the monsoon. They do not breed in still waters of ponds. They only mature there. This is a traditional method of collecting spawn, the fishseed.

(m)

The seed collected like this often consists of high percentage of unwanted species. At this stage, the desirable varieties cannot be separated from the undesirable ones. Valuable space of ponds is wasted by raising undesirable seed and the result is the poor yield.

(m)

Transporting the spawn for marketing and distribution is often unscientific and is the cause of heavy mortality.

(m)

How can we increase the survival and growth of fish and get increased production to provide more protein to our people? This film will show you how this can be done by INDUCED BREEDING.

(m)

At the substation of the Central Inland Fisheries Research Institute at Cuttack, scientists have evolved a technique of induced breeding of cultivated species of fish by administration of fish pituitary hormones. For induced breeding the first step is to procure fully ripe fresh fish for the collection of pituitary glands.

(m)

The most appropriate time for this is the breeding season. Success in induced breeding depends to a great extent on the potency of the gland material collected. Sufficient care is therefore taken in selecting the most potent donor fish for this purpose.

(m)

The glands are removed by dissecting the heads of the donor fish taking sufficient precautions to see that they are not damaged.

The pituitary gland in fish is a small body situated below the brain in a cavity on the floor of the brain box. This gland secretes a large number of hormones of which the gonad-stimulating hormones are mainly responsible for the maturing of the reproductive cells and the spawning of fish.

THE PITUITARY GLAND

Immediately after removal the glands are immersed in absolute alcohol.

(m)

Absolute alcohol dehydrates and defattens the glands.

(m)

The preserved glands are taken out for making extracts.

(m)

The filter paper absorbs the absolute alcohol.

(m)

After recording the weights, the glands are macerated in a little distilled water.

(m)

Lather on the extract is diluted to a measured volume of distilled water.

(m)

The extract is then centrifuged. With centrifugal action the tissue particles settle down at the bottom. The supernatant fluid containing the hormones is then decanted out for injection. For major induced breeding operations extracts of large quantities of glands are made. These are preserved in glycerine and ampouled for future use.

These hormone ampoules are distributed to various fish breeding centers of the State Fisheries Departments, Agricultural Universities and also to the fishfarmers.

The technique is so standardized and simplified that even an ordinary fish farmer is able to breed fish.

(m)

Now for feeding a breeder fish .... Raising a brood stock of healthy breeder fish is essential for successful induced fish breeding. They are raised in special ponds and are regularly fed with artificial feed such as rice bran and oilcake.

(m)

With the advent of monsoon when the water temperature goes down, induced breeding operations start. Breeders are netted and healthy ones are selected.

(m)

How to distinguish a mate from a female?

A male is recognized by the roughness of the upper surface of the pectoral fin in contrast to that of a female which is smooth:

Now the selection of proper breeders. This is very important. A fully ripe male freely oozes milt when the abdomen is gently pressed. A fully ripe female will have a swollen pinkish genital opening and a soft bulging rounded abdomen.

(m)

Weights of the selected breeders are then taken. They are now ready for hormone injections.

(m)

Proper dose of pituitary hormones is determined according to the weight of the breeder. The female is usually given a preliminary low dose of 2 of 3 mg of pituitary gland per kg of its body weight. After about 6 hours a second dose of 5 to 8 mg of glands per kg body weight of the female is given. The male breeder receives a single dose of 2-3 mg per kg body weight at the time of the second injection to the female. An intramuscular injection of the selected breeders.

The breeding hapa or the Honey-moon apartment.

It is a cloth container with a cover fixed on bamboo poles and submerged in pond water to three-fourths of its height.

(m)

One set of breeders comprising two males and one female is introduced into one breeding hapa. This is done usually in the evening.

(m)

After the introduction, the mouth of the hapa is securedly closed so that they do not escape the honey-moon.

(m)

The injected hormones stimulate the ovaries to release the eggs, and the testes to release the sperms within 3 to 6 hours of introduction at water temperature between 24 to 31°C. The fertilization of the eggs by sperms taken place in water.

(m)

In the morning the breeders are removed from the hapa and the hapa containing eggs with developing embryo is taken out and the eggs are collected for hatching. Fertilized eggs absorbed water and have swollen to the size of peas. A quantitative estimation of the eggs is made before distribution for hatching.

From a single female breeder weighing 4 kg we may expect 1.0 to 1.2 million eggs.

(m)

This is the method used for induced breeding so far as popular Indian carps such as, catla, rohu and mrigal are concerned. But for the recently introduced Chinese varieties like grass and silver carps, the method of induced breeding is slightly different. For selecting a ripe female for breeding, a catheter is inserted through the genital opening and a few eggs removed from the ovaries for testing their stage of ripeness.

(m)

About 3 to 4 h after the second injection to the female, it is examined and as soon as the female is ready to release eggs, it is carefully taken out and eggs stripped on a basin. The male is then similarly taken out and stripped and milt spread over the eggs.

Simultaneous mixing of the eggs with sperms with a feather is done so that fertilization takes place in the basin.

(m)

Shortly afterwards, pond water is added and the excess of milt is washed out. Fertilized eggs start swelling.

(m)

Eggs of different species kept separately, are now ready for going to the hatchery.

(m)

The Hatching Hapa

It consists of two cloth containers. The outer container is made of close-mashed fine cloth so that the hatchlings do not escape from it. But the smaller inner container is made of round-meshed mosquito net cloth.

(m)

The eggs are introduced inside the inner container. About 100,000 eggs are distributed evenly in each hapa.

(m)

The eggs hatch out within 15 to 18 h after fertilization at water temperature of 27° to 31°C. The fully developed embryo now starts making efforts to wiggle out of its shell.

(m)

The eggs having hatched, the hatchlings escape into the outer hapa through the meshes of the inner hapa. Inner haha containing only the egg shells and bad eggs is removed leaving all the hatchlings in the outer hapa.

An outdoor hatchery has several drawbacks; very often heavy mortality occurs due to natural hazards. The hatching of eggs can as well be done indoor. The Cuttack substation of CIFRI has successfully developed a glass jar hatchery with continuous circulation and oxygenation of water under controlled temperature. About 50,000 fertilized eggs are put in each jar. The eggs are under constant movement. The flow of water through the jar is so regulated that the eggs do not flow out. After the eggs hatch out, the newly hatched hatchlings have a natural tendency to go up. Only the hatchlings therefore, escape out of the jars along with the flow of water.

Research carried out at CIFRS Cuttack has also demonstrated that by following the improved techniques evolved by them, the survival and growth of fish can be increased and fish production enhanced.

(m)

Finally, all the hatchling collected in the cloth container known as the receptacle.

(m)

Back to the outdoor hatching hapa where the hatchlings are growing, yolk is getting slowly absorbed and the mouth is being formed.

(m)

On the third day after hatching the mouth is formed and the hatchling start feeding. The baby fish is now called young fry or spawn. The spawn is carefully collected from the outer hatching hapa and the quantity estimated.

(m)

For a long distance journey the spawn is put in plastic bag and is then packed with oxygen under pressure and transported to the pond site.

This ends the story of the induced breeding of fish. This dependable technique of production of pure fish seed has revolutionized the traditional practice. If offers vast scope for development of fish culture, which to the people of India means a good source of protein rich food.

Reference Books on Aquaculture

1. Aquaculture : The Farming and Husbandry of Freshwater and Marine Organisms – by J. Bardach, J. Ryther and W. Mclarney (1972).
2. Fish Culture – by C.F. Hickling (1971) Faber and Faber, UK.
3. Textbook of Fish Culture : Breeding and Cultivation of Fish – by Marcel Huet.
4. Freshwater Fishpond Culture and Management – by M. Chakroff.
5. Freshwater Fisheries and Aquaculture in China – by D.D. Tapiador *et al.* (1977).
6. Elementary Guide to Fish Culture in Nepal – by E. Woynarovich.
7. Fish Culture in Central East Africa – by A. Marr *et al.* (1966).
8. Advances in Aquaculture – edited by T.V.R. Pillay and W.A. Dill.
9. Advances in Fish Sciences and Technology – by J.J. Connel (Editor).
10. Aquaculture Practices in Taiwan – by T.P. Chen.
11. Modern Aquaculture for the Philippines – by C. delos Santos Jr.
12. World Fish Farming : Cultivation and Economics – by E. Brown.
13. Better Freshwater Fish Farming : The Pond – FAO 1981.
14. Profitable Cage Culture – by G. Nett & P. Barrett.
15. Raising Fresh Fish in Your Home Water – by B. Bertz and M. Podems.
16. Principles and Practices of Pond Aquaculture – edited by Lannan *et al.*

17. Textbook of Pond Culture – by W. Schaeperclaus Fish. Leaflet 311 USDI.
18. Development of Aquaculture in the Philippines – by Schmittou *et al.*, Auburn.
19. Aquaculture – by P.J. Reay.
20. Water Quality Management for Pond Fish Culture – by C.E. Boyd.
21. Bioeconomics of Aquaculture – by P.G. Allen *et al.*
22. Aquaculture in Japan – by Aiko Homma.
23. Modern Methods of Aquaculture in Japan – by T. Kafuku and H. Ikenoue (Editors).
24. Farming Marine Fishes and Shrimps – by P. Korringa.
25. Giant Prawn Farming – by M.B. New (Editor).
26. Genetics in Aquaculture – by N.P. Wilkins and E.M. Gosling.
27. Channel Catfish Culture – by C.S. Tucker.
28. Fish and Fisheries of India – by V.G. Jhingran.
29. Planning of Aquaculture Development – by. T.V.R. Pillay.
30. The Farming of Fish – by C.F. Hickling.
31. Fish and Shellfish Farming in Coastal Waters – by P.H. Milne.
32. Aquaculture : Farming in Water – by L. Wolfe.
33. Pond Construction and Management – by Alfredo M. Lopez.
34. Farming the Edge of the Sea – by E.S. Iversen.
35. The Philippines Recommends Bangos – by PCARRD.
36. The Philippines Recommends Tilapia – by PCARRD.
37. Handbook of Fish Culture in the Indo-Pacific Region – by S.L. Hora.
38. Integrated Agriculture – Aquaculture Farming Systems – by Pullin/Shehadeh.
39. Proc. of the First International Conference on Culture of Penaeid Prawns/Shrimps – edited by Y. Taki *et al.*

40. Fish Farming Handbook – by E.E. Brown and J.B. Gratzek.

41. Freshwater Prawn Farming : A Manual for the Culture of *M. rosenbergii*; FAO.

42. Tilapia : A guide to their biology and culture in Africa (1979) – by Balarin.

43. Biology and Culture of Tilapia (1982) – edited by R.S. Pullin and W.J. McConnell.

44. Coastal Aquaculture in the Indo-Pacific Region – edited by T.V.R. Pillay.

45. Principles of Warmwater Aquaculture – by Robert R. Stickney.

46. Commercial Fish Farming (1981) – by Hepher & Y. Pruginin.

PART TWO

LIMNOLOGY

1

General Limnology

Limnology: The word originated from two Greek words : **'Limnos'** : means lakes, pools and swamps and **'Logos'** : means discourse or study.

The term **Limnology,** therefore generally means; study of lakes.

Recent Definition (Goldman and Horne) :

"Limnology is the study of fresh or saline water which are contained within continental boundaries."

More Elaborate Definition (Wetzel) :

"Limnology is, in broad terms–the study of the function relationships and productivity of biotic communities as they are affected by the dynamics of physical, chemical and biotic environmental parameters."

The well-known Limnologist Welch, however, suggested another definition :

"Limnology is now commonly defined as that branch of science which deals with **biological productivity** of **inland waters** and with all the **casual influences** which determine it."

What is Biological Productivity?

Biological Productivity of any body of water is the end result of the interaction of the organisms present, with the surrounding environment.

Inland waters include all types of waters such as, running (lotic) or standing (lentic); fresh, salt or brackish which are wholly or almost completely included within the land masses.

Casual influences involve physical, chemical, biological, meteorological and others, which determine the character and quantity production.

Limnology depends upon proper application of certain facts, principles and methods of Physics, Chemistry, Geology, Hydrograpy, etc. But biological productivity is the central unifying feature which ties the whole subject into a coherent, orderly organized field.

2

OUR PLANET EARTH

Earth

Lithosphere (Comprises Landmasses)

Hydrosphere (includes, oceans, lakes, streams and sub-terranean waters.)

Table 1 : Water in the Biosphere

	Volume (1000 km^3)	*Per cent of total*	*Renewal time*
Oceans	1,370,000	97.61	37,000 years
Polar ice glaciers	29,000	2.08	16,000 years
Groundwater (activity exchanged)	4,000	0.29	300 years
Fresh Water Lakes	125	0.009	1-100 years
Saline Lakes	104	0.008	10-1,000 years
Soil and subsoil moisture	67	0.005	280 days
Rivers	1.2	0.00009	12-20 days
Atmospheric Water Vapour	14	0.009	9 days

(By j. Valleratyne : from Watzel)

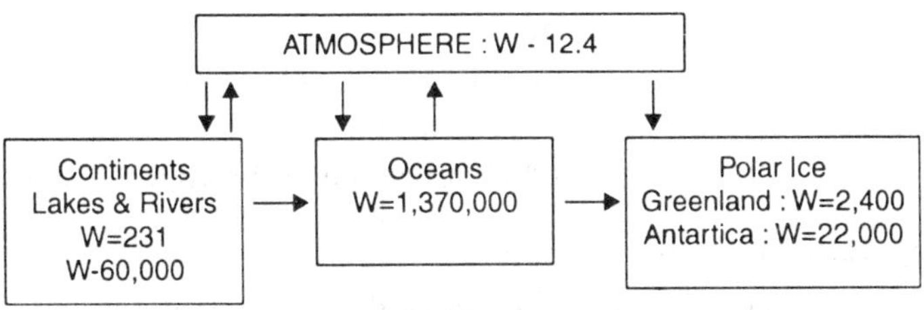

Fig. 1 : Global Water Balance

W=Water content in $10^3 km^3$. Values in Arrows : Transport in $10^3 km^3\ yr.^{-1}$

*Estimate of groundwater is to depth of 5 km of Earth's crust (Wetzel).

3

WATER BALANCE IN LAKES

Sources of Waters in Lakes

1. Precipitation falling on the lake surface.
2. Water in surface influents.
3. Groundwater seeping through the floor of the lake.
4. Groundwater entering through springs.

Loss of Water from the Lakes

1. Outflow from surface.
2. Seepage to groundwater.
3. Evaporation from surface.
4. Transpiration.

Types of Lakes according to Water Losses

1. *Drainage lakes* : In these lakes, loss of water occurs by flow from an outlet, such as a river or steams.
2. *Seepage lakes* : Loss of water occurs through seepage into the groundwater through the basin walls.
3. *Closed lakes* : The lakes have no outflow effluents and loss water only by evaporation.
4. *Open lakes* : Those lakes that have an outflow by an outlet or loss water through seepage.

Hydrological Cycle

The hydrological cycle consists of three principal phases :

1. Evaporation
2. Precipitation and

3. Groundwater run off.

Each phase involves transport, temporary storage, and a change in the state of the water.

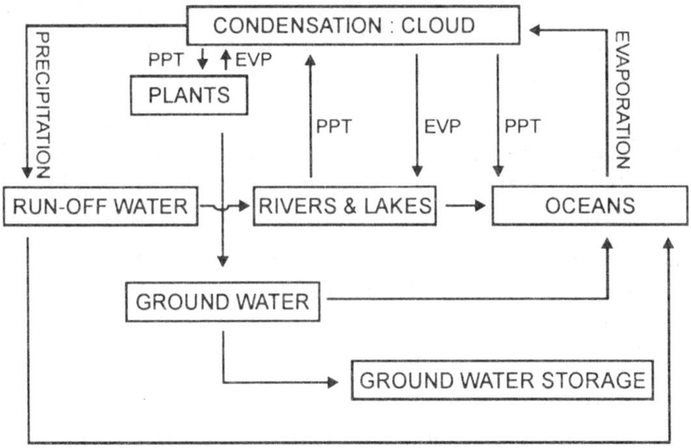

Fig. 2 : Diagram showing phases of Hydrological Phases

Lakes, their Origin, Forms and Distribution

Definition of a Lake :

"Lakes are inland bodies of water that occupy depressions in the surface of the land. The depressions are called **basins.**"

OR

A lake is sometimes defined as a body of standing water completely isolated from the sea and having an area of open, relatively deeper water sufficiently large to produce somewhere on its periphery a barren wave-swept shore (Welch).

Lake Basins are formed by :

I. **Tectonic Agent** : Tectonic basins are depressions formed by movements of deeper portions of the earth's curst.

A. Those formed by faulting of the Earth's crust :

(i) Basins associated with **single fault :** e.g. Abert L. (Oregon).

(ii) Basins associated with **multiple faults,** known as **Grabens :** e.g. Lake Tahoe (Cal., USA), L. Baikal (Siberia),

L. Biwa (Japan) and also the chain of rift lakes in Africa e.g. L. Malawi, L. Tanganvika and L. Mobutu.

B. Those arising from the uplift of portions of the seafloor ; e.g. Lake Okeechobee (1840 km^2), Florida, USA; Caspian and Aral Lakes (Saline) were also created by uplift of the sea bottom or closing of the Tethys Sea by Plate Tectonics. These basins, called Relict lakes retain their original marine structures.

C. Those formed by upwarding of Earth's crust so as to cause the reversal of the existing drainage system ; e.g. L. Koga (formed by the reversed flow of Kafu River in E. Africa.)

D. Those formed by uplifting of the land surrounding the basin : e.g. Lake Victoria (E. Africa), Lake Triticaca in the Andes (Peru).

II. **Volcanic Origin**: Basins created by volcanoes.

A. *Crater lakes* : These formed when a crater of an extinct volcano remains intact and later on gets filled with water. They are of 3 types.

(i) Those existing in the unaltered cones of previous volcano : e.g. L. Crater Butte (California).

(ii) Those formed on the depression resulting from abortive volcanic explosion. These lakes are formed at considerable depths below the earth's surface resulting in low rims but deep basin. These are known as **Maars** (diameter<2 km but even more than 100m deep). e.g. Pulvermaar in Germany (0.35 km^2; 74 m deep).

(iii) Those resulting from the collapse of the rim of a volcano. Collapse cones which form the largest basins are known as **Calderas** e.g. Crater lake, Oregon (10km/600m deep).

B. Basins created by damming the outlet of a valley by lava flow; e.g. Snag lake, California.

III. **Glacial Origin** : Lakes formed by glacial erosion.

A. Glacial rock basins formed by gouging (digging) action of a moving glacier. These are two types :

(i) Large valley rock basins formed by continental ice e.g. Great Slave L. (Canada); Great Laurentian lakes of N. America (L. Superior, L. Michigan, L. Huron, L. Erie and L. Ontario).

(ii) Lakes which are formed at about the snow line in glaciated valley. These lakes are semicircular and are known as **Cirque Lakes,** e.g. Castle Lake, California.

B. *Draft Basins* : Lakes formed by melting of huge blocks of ice trapped in glacial morainee. These are known **Kettle Lakes,** e.g. Walden Pond (Massachusatts).

(Debris carried down and deposited by glaciers are termed Moraines).

C. lakes formed by the deposition of Moraines, e.g. Green L. (Winconsin).

D. Lakes formed by ice or by Moraine in contact with existing ice, e.g. Gapsban L. (Central Asia).

IV. Lakes with Miscellaneous Origin

A. **Basins formed by landslides** : Due to sudden movement of large quantities of unconsolidated materials in the form of landslides into the floors of a river valley (occasionally initiated by the activity of earthquakes), e.g. Lake Buhi (Philippines), L. Sarez, Pamir Mountain.

B. **Solution lakes** : Lakes formed in depressions due to solution of limestone by water and also by dissolution of other soluble rocks : e.g. Deep Lake (Florida).

C. **Fluviatile - dam lakes** : These lakes are created by the deposition of sediments by a tributary across the main river, e.g. Tulare Lake (California).

D. **Lakes formed by flood plains of rivers** : Oxbow lakes are formed when loop of **Meander** (S shaped winding river) is cut off by silt deposition, e.g. Oxbow lakes in the flood plains of most rivers of the world.

E. **Man-made lakes** : (artificial man-made reservoirs). A lake is formed by construction of a dam built across a river valley thereby blocking the flow of water : e.g. Hover Dam (R. Colorado), Aswan Dam (R. Nile in Egypt) and thousands of other dams.

F. **Lake basins formed by wind section** : Wind action effects formation of lakes by deflation or erosion of broken rock or by redistribution of sand; the resulting lakes are known as **Dune Lakes** : e.g. Lake Paoay (Philippines), Moses L. (Washington).

G. **Lakes formed by shoreline activity** : These lakes are formed by building up of a sandbar across some irregularities of the coast line, e.g. Lakes of Landes (West coast of France).

H. **Lakes of Organic Origin**

(i) Damming due to organic accumulation (deposition of decomposed plants and associated detritus), e.g. Silver L. (Halifax).

(ii) *Beaver Dams* : A large number of very small lakes in N. America have been formed partly or wholly by the activities of beavers.

4
SOME IMPORTANT LAKES OF THE WORLD

Table 2

Name	*Continent (Country)*	*Area km³*	*Depth Zm m*	*z m*	*Type of Basins*	*Remarks*
Superior	N. America (USA-Canada)	83,000	307	145	Glacial	Largest freshwater
Michigan	U.S.A.	57,580	265	99	Glacial	lake. The
Huron	U.S.A.-Canada	59,510	223	76	Glacial	Laurentian
Erie	U.S.A.-Canada	25,820	64	21	Glacial	Great Lakes
Ontario	U.S.A.-Canada	18,760	225	91	Glacial	(Area : 242,000 km² constitute the greatest continuous mass of fresh water
Great Slave	Canada	30,000	614	–	Glacial	
Great Slave	Canada	29,000	137	–	Glacial	
Tanganyika	Africa	34,000	1,470	572	Tectonic (Graben)	Second deepest in the world
Malawi (Nyassa)	Africa	30,000	706	203	Tectonic (Graben)	
Victoria	Africa	68,000	79	40	Tectonic (up-warding)	Wet season : 40,000 km²
Chad	Africa	16,000±	12±	1.5±	Tectonic	Dry season : 6,000 km²
Baikal	Asia (USSR)	31,550	1,741	730	Tectonic (Graben)	Deepest lake in the world. 20% all fresh-waters
Balkhas	USSR	17,575	26.5	6.13	Tectonic	
Caspian Sea	Asia	436,400	946 182		-do- uplift of sea bottom	Largest Lake (Saline)
Aral Sea	Asia (USSR)	62,000	68	15.6	-do-	Saline

Some Important Lakes of the World

Name	*Location*	*Area*	*Zm m*	*Origin*	*Remarks*
Laguna	Luzon (Laguna)	900 km^2	2.5m	Tectonic	
San Pablo Lakes	Laguna	2.9 km^2 (Total)	7.2-156m	Volcanic (Caldera)	Consist of Seven Lakes
Tadlak	Laguna	0.27 km^2	65m	-do-	
Taal	Luzon (Batangas)	263.65km^2	198m	Volcanic	
Buhi	Luzon (Camarines Sur)	10 km^2	10m	By landslides	
Paoay	Luzon (Ilocos Norte)	4.4 km^2	4.1m	Dune Lake	
Mainit	Mindanao (Surigao del) Norte)	141 km^2	223m	Tectonic	
Naujan	Mindoro (Oriental)	79 km^2	19m	Volcanic	
Lanao	Mindanao	357 km^2	112m	Volcanic	(lava flow)

1. Lake Bunot 2. Lake Calibato 3. Mohicup
4. Lake Palacpequen 5. Lake Pandin 6. Sampaloc Lake
7. Lake Yambo

5

GEOMORPHOLOGY OF LAKE BASINS

How does this study help study of lakes?

1. It reflects physical, chemical and biological events in the lake basins.
2. It plays a major role in lake's metabolism and its control.
3. It controls the nature of lake's drainage.
4. It controls inputs of nutrients in the lake.
5. It regulates volume of influx in relation to flushing-renewal time.

Shape of a lake basin in reflected in its Productivity

a. steep-sided 'U' - or 'V' - shaped basin (usually formed by tectonic forces) are ordinarily deep and unproductive.

b. shallow basins with greater percentage contact of water with soil show higher productivity.

c. basins with longer, irregular shorelines are usually more productive because of greater contacts of water with soil.

Form of shorelines in relation to lake origin

a. **Circular** : Crater lakes, Calderas & Maars; doline or solution lakes. Development of shoreline approaches unity in such lakes. (1.04-1.15 in Maars)

b. **Sub-circular** : Cirque lakes, kettle lakes

c. **Elliptical** : Oriented lakes of Arctic

d. **Subrectangular Elongate** : Most of the Grabens

e. **Dendritic** : When a high dam is formed flooding the tributary valleys (L. Kioga)

f. **Lunate** : Oxbow lakes; also a few special type of calderas.

g. **Triangular** : Lakes in drowned valleys behind bars; shoreline lakes

h. **Irregular** : Regions of glacial scouring and where several basins have been fused to form a lake.

Insulosity

Insulosity of a lake is the percentage of area within the shoreline that is occupied by island. L.Huron has the largest island inside a freshwater area.

Morphometry of Lakes

A. Bathymetric Map (Recording of Lake Morphometry) : Construction

At first a good topographic map or an accurate outline map of the lake to be prepared. Then from this Bathymetric map can be done in the following three ways :

1. Fixing position at which soundings are taken on or from the shore.
2. Use of ordinary land survey methods.
3. Counting of oar strokes on a course across the lake.

B. Morphometric Parameters

1. Length (1) : Shortest distance, through the water or on the water surface, between the most distant points on the lakeshores.
2. Breadth (bx) : The length of a line from shore to shore cutting the line, defining length at right angles at any point x.
3. Mean breadth ($\overline{b}$) : is equal to the area divided by

   Length = $\overline{b} = \frac{A}{1}$
4. Maximum depth (Zm) : The greatest depth of the lake.

5. Mean depth ($\bar{Z}$) : Dividing *volume* of the lake by its area $\bar{Z} = V/A_0$

6. Area (A) : It is determined by planimetry from the outline of the map (surface area). The area of contour of depth Z is A_z.

7. Volume (V) : is the integral of the areas of each stratum at successive depths from the surface to the point of maximum depth. Volume can be estimated by summation of a series of truncated cones of the strata.

$$V = \frac{h}{3}(A_1 + A_2 + \sqrt{A_1 A_2})$$

8. Shore line (L) : Length of shore line is determined directly or from maps with a map measurer (chartometer or ratometer).

9. Shore line development (D_L) : the ratio of the length of the shore line (L) to the length of the circumference of a circle of area equal to that of the lake.

$$D_L = \frac{L}{2\sqrt{\pi A_0}}$$

6

LIGHT : OPTICAL PROPERTIES OF LAKES

Light exerts a very good profound influence upon a whole series of biological phenomenon in lake waters.

What is light?

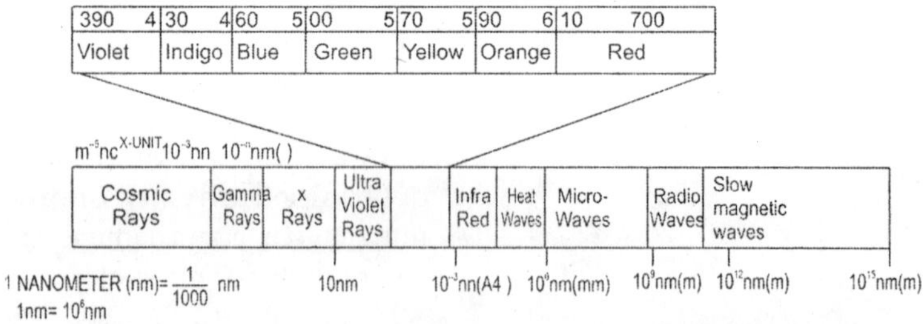

Fig. 3 : The electro-magnetic spectrum with wave lengths in Nannometer (nm).

Visible spectrum : 380-760 nm wave lengths

Ultra-violet rays : shorter wave length; (but higher frequency)

Infra-red rays : longer wave length; but lower frequency

Photosynthesis occurs within 400-800 nm wave lengths.

Of the electro-magnetic spectrum, the narrow band between wave lengths 380-760 nm which is sensitive to human retina is termed **Light.** In very intense artificial sources, the extreme limit is between 310-1050 nm.

Quanta or Photons

Light reaching the earth can be thought of as a continuous flow of electro-magnetic waves or alternately, as **Photons** or **Quanta.** Which are discrete packets of energy. Wavelength is useful measure of light colour while quanta most conveniently define energy. **Light intensity** means the number of **quanta** passing through a certain unit area. **Light energy** is measured in Einteins. One Einstein is equivalent to 2.854 × 10^7 gram-calories.

Light is an Energy

Almost all energy that controls the metabolism of lakes is derived directly from the solar energy utilized **Photosynthesis.**

Solar energy that reaches the surface water of a lake determines the intesity and quality of light and also affects the temperature of water through radiation, evaporation and conduction. Solar radiation incident on the lake surface is both **direct** sunlight and **indirect** or diffuse light which is scattered and reflected of mountains, clouds and atmospheric particles.

A beam of light falling on a water surface at a certain angle (angle of incidental) is **in part** penetrates the water and **in part** reflected. The proportion of reflection depends on the angle of incidence.

Light penetrating water is first refracted (water is transparent liquid), then either **absorbed, scattered or** transmitted. Light that is **absorbed** is changed to heat and the intensity of light decreases.

Factors influencing Light Generation

(a) intensity at surface

(b) angle of incidence of light

(c) difference in lattitude

(d) seasonal differences

(e) diurnal differences

(f) dissolved materials and

(g) suspended materials.

The amount of light that is reflected from lake surface on a clear **summer** day is 5 to 6% (average). This value increases in **Winter.** the percentage of reflected light is called **Albedo.**

Scattering of light from the water results in the loss of large amounts of light from lake. It changes significantly with : (a) depth, (b) seasons, (c) location of lake and (d) variations in distribution of particulate matter.

Absorption and Transmission of Light in Water

Absorption of light in water is defined as diminution of light energy with depth by transformation to heat. The percentage held back in 1 m is termed, "Percentile absorption". The percentage that is transmitted is called, "Percentile transmission" and is expressed by the formula :

$$\frac{100\,(I_0 - I_z)}{I_0}$$

where I_0 = irradiance at lake surface

I_z = irradinace at depth z (here 1 m)

However, the intensity of light $I_z = I_{0e}$ — nz)

where n= Extinction coefficient (constant for a given wave length).

Solar radiation entering pure water is absorbed selectively. Minimum absorption is at 470 nm (Blue); maximum is at 750 nm (Red). Eighty per cent is absoprtion than pure water.

Measurement of Light under Water

By Secchi's Disc (Simplest Optical Measurement)

Secchi-Disc transparency is determined by the mean depth of the point where a weighted 20 cm diameter white disc disappears and the point where it reappears after raising it, after it has been lowered beyond visibility.

Colour of Lakes

Lake colour is composed of reflected light consisting of **Real** and **Apparant** colour, sometimes complicated by bottom colour in shallow lakes, 'Real' as oposed "Apparent colour depends on the interaction between the wavelengths that are scattered back and absorption of these wavelengths in water between the depth of scattering and the surface.

Deep lakes **without or with very little** dissolved or suspended matters appear **Blue.** Under stormy cloud they appear **Grey** or **Brown.** Glacial valley lakes (nearest to glaciers) look **White** due to suspended glacial flows. Lake with large quantities of dissolved materials (especially organic) appear **Yellow** or **Brown.**

Phytoplankton blooms produce **Green** colour. Spectacular lake colours (Red or Blue-green) are produced by bloom of near surface blue-green algae.

Turbidity

Turbidity is a condition of water resulting from the presence of suspended matters, which may be of **Biogenic** (living) or **Abiogenic** (non-living) origin.

(a) **Settling suspended matter** : These are substances which will sooner or later settle down to the bottom in still water. Rate of settling depends on the specific gravity, size and by the ratio of surface to volume of the particles and by the viscocity of water.

(b) **Non-settling suspended matter :** These are either very finely divided solids or those with specific gravity less than water and are in permanent suspension.

(i) Planktonic organisms and coarsely divided non-living substances whose specific gravity is such that they are constantly suspended.

(ii) Very finely divided non-living materials and orgnisms of exceedingly small size, such as nano plankton.

7

LACUSTRINE ZONATION

Zonation

1.	Epilittoral Zone	: Lies entirely above water level uninfluenced by the spray of waves.
2.	Supralittoral Zone	: Also lies entirely above the water level but is subject to spraying by waves.
3.	Eulittoral Zone	: Shore zone between the highest and lowest seasonal water levels and is often influenced by the disturbance of breaking waves.
3(a).	Infra or sublittoral zone	: The shoreline region lying between the lowest water level to the lower boundary of macrophytic growth.

This (3a) is divided into three sub-zones. They are :

(i).	Upper infra-littoral zone	: Zone emergent rooted vegetation.
(ii).	Middle infra-littoral zone	: Zone of floating-leaved rooted vegetation.
(iii).	Lower infra-littoral zone	: Zone submerged rooted or adnate macrophytes.
(iv).	Littori profoundal zone (transitional zone)	: Zone scattered photosynthetic forms and includes a gradient of benthic algae, manly blue-green green algae.
(v).	Profoundal zone	: Zone comprising of exposed fine sediment free of vegetation.
(vi).	Pelagic zone	: Free open water of lakes.

8
CIRCULATION OF FOOD IN A LAKE

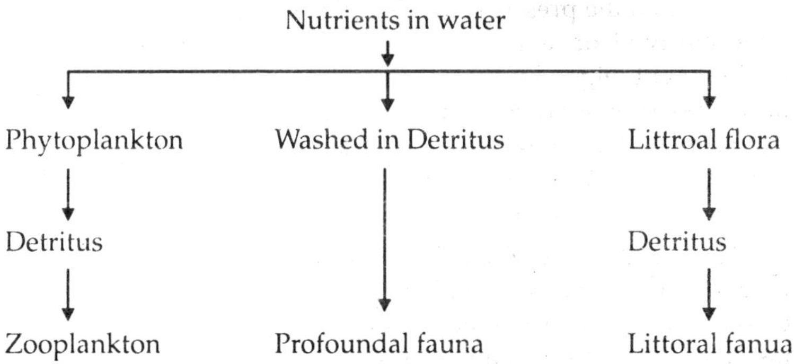

Fish Production

In considering circulation of food materials in a typical lake the following fundamental facts must be kept in mind :

A. The ultimate basic substances are :

(a) Inorganic nutrients dissolved in water and

(b) Certain energy and gases from the atmosphere.

B. Only the chlorophyll-bearing littoral flora and phytoplankters, and certain bacteria can utilize directly these basic materials in constructing living matter.

C. All other organims, plant or animal, rest as a dependent superstructures upon those mentioned in item (B).

D. Every organism in lake population may :

1. by death and destruction contribute directly to the dissolve materials and detritus OR

2. be consumed as food by other organisms.

9

WATER AND LIFE

The evolution of complex biochemical systems based on carbon depended on the presence of water. Without water, life could not have evolved on earth, no matter how suitable the earth's environment might have been in other respects. Penciful systems of life based on ammonia and silicates rather than on water and carbon have been suggested, but such speculations are totally hypothetical and in any case would require environmental conditions vastly different from those on earth.

Water is the most abundant of all compound in cells, making up about 60% to 90% of most living organisms. The maintenance of a constant aqueous internal environment is a major physiological task for all organisms, both terrestrial and aquatic.

Water has several extraordinary properties that makes it especially fit for its essential role in living systems. It is the most stable yet versatile of all solvents. Water is the only substance that occurs in nature in the three places of solid, liquid and vapour within the ordinary range of earth's temperature. Water has a **high specific heat capacity** : 1 calorie is required to elevate the temperature of 1 g of water 1°C (such as from 15°C to 16°C). Every other liquid but ammonia requires less heat accomplish the same temperature increase. Water has twice the specific heat capacity of alcohol or oil, four times the specific heat capacity of aluminium, and 10 times the specific heat capacity of iron. Thus, the high thermal capacity of water has a great moderating effect on environmental temperature changes and is a great protective agent for all life.

Water also has a **high heat of vapourization.** More than 500 calories is required to change 1 g of liquid of water into water vapour. This is twice as much heat as is required to vapourize methyl alcohol. For terrestrial animals (and plants), cooling produced by the evaporation of water is an important means of getting rid of excess heat.

Another important property of water from a biological standpoint is its **unique density behaviour** during changes of temperature. Most liquids become denser with decreasing of temperature. Water, however, reaches its maximum density at 4°C **while still a liquid,** and then becomes less dense with further cooling. Therefore ice **floats** rather than forming on the bottom of lakes and ponds. If it were not for this property, bodies of water would freeze solid from the bottom up in winter, and except in warmer climates, would not necessarily completely melt in summer. Under these conditions, aquatic life would be severely limited.

Water has a **high surface tension,** greater than that of any other liquid mercury. This property, caused by the great cohesiveness of water molecules, is important in the maintenance of protoplasmic form and movement. It also creates a unique ecological niche for insect forms, such as water striders and water boatmen, that skate on the surfaces of ponds. Despite its high surface tension, water has **low viscosity,** and property that favours the movement of blood through minute capillaries and of cytoplasm inside cellular boundaries.

From what we have siad, clearly it is the stability of water that makes it such a fine medium for live. Actually, water's great resistance to change in state is surprising because water is composed of oxygen and hydrogen, two of the most reactive elements known. We now know that the remarkable properties of water can be explained in large part on the basis of its structure. Water is hydride of oxygen. The water molecule is shaped like an isosceles triangle; the legs of the triangle are two O-H bonds that subtend an angle of approximately 105 degrees (Fig. 5).

The water molecule is electrically neutral because it has an equal number of protons and electrons. However, the oxygen is so strongly electrognegative that it tends to draw electrons away from the hydrogen nuclei. Consequently, each of the hydrogen atoms carries a net charge that is slightly positive (ε +), and the oxygen atom carries a net charge that is slightly negative (ε −) and twice as great as the hydrogen net charge. Thus, the water molecule is a permanent dipole with zones of negative and positive charge. Some sharing of the bonding pairs of electrons continues between the hydrogen and oxygen nuclei. Therefore, these charges are not complete ionic charges (+ and —), but rather partial changes (δ +

and ε −) with the charge on each oxygen atom twice as great as that on each hydrogen atom.

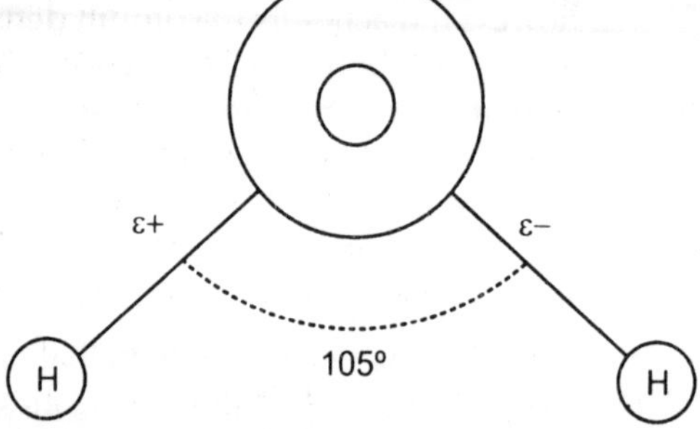

Fig. 4 : Molecular structure of water. The 2 hydrogen atoms bonded covalently to an oxygen atom are arranged at an angle of about 105 degrees. Since the electrical charge is not symmetrical, the molecule is polar with positively charged and negatively charged ends.

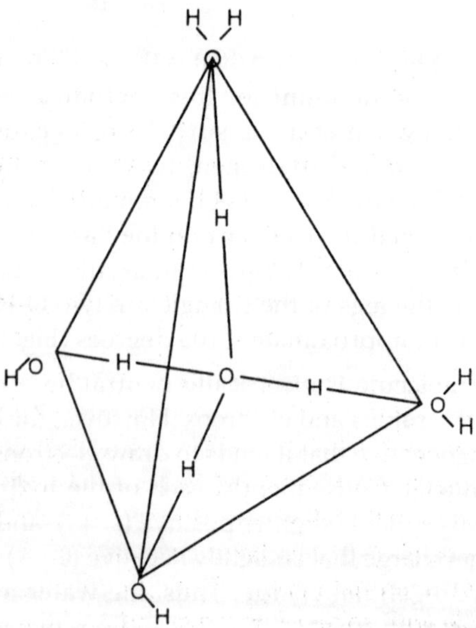

Fig. 5 : Geometry of water molecules. Each water molecule linked by hydrogen bonds (dashed lines) to 4 other water molecules. If imaginary lines are used to connect the divergent oxygen atoms, a tetrahedron is obtained. In Ice, the individual tetrahedrons associate to form an open lattice structure.

In ice, water molecules tend to orient themselves so that the positively charged zone of one molecule faces the negatively charged zone of another. Each oxygen atom therefore becomes the centre of a tetrahedron of other oxygen atoms (Fig. 5). The kind of linkage indicated in the diagram by a dashed line is a hydrogen bond. Although it is much weaker than the O—H bonds within each water molecule, it nevertheless requires substantial energy for breakage (approximately 5 kcal/mole). Hydrogen bonds are of great importance in the chemistry of life because they serve as atomic bridges that participate in the formation of macromolecules, especially proteins.

The tetrahedral arrays of water molecules in ice form an extensive, open, crystal-like network held together by hydrogen bonds. When ice melts, about 15% of the hydrogen bonds break, but in its usual liquid state, water is still thought to possess a fair degree of order. In fact, water molecules are more densely packed together in liquid water than in ice, and this is why ice floats rather than sinking to the bottom.

Hydrogen bonds gives water most of its special properties. The attraction of water molecules to each other through hydrogen bonds is responsible for the high surface tension of water. The high specific heat capacity of water is consequences of attractive force between hydrogen bonds. When water is heated, much of the heat energy is used to rupture some the hydrogen bonds, leaving a relatively small amount of heat to increase the kinetic energy (molecular movement), and thus the temperature, of the water. Similarly, the heat of vapourization of water is high because all of the hydrogen bonds between a water molecule and its neighbours must be ruptured before water can escape the surface and enter the air.

Finally, water is an excellent **solvent** for the ions of salts, which are so important to life processes. Salts dissolve to a much greater degree in water than they do in any other solvent. This results from the dipolar nature of water, which causes it to orient itself around charged particles dissolved in it. When, for example, NaCl is dissolved in the water, Na^+ and Cl^- ions present in the solid salt rapidly separate into independent Na^+ and Cl^- ions. The negative zone of the water dipoles align themselves around the Cl^- ions. This keeps the ions separated, promoting a high degree of dissolation. Other solvents having less or no dipolar character are less able to align effectively around such ions and are, therefore, less able to dissolve the salt.

Water and life are part and parcel of each other. The special conditions on earth resulting from its ideal size, element composition, and nearly circular orbit at a perfect distance from a long-lived star, the sun, made possible the accumulaton of water on the earth's surface. It is difficult even to imaginze the origin of life without water.

10
Water as an Environment

Water has two-fold effect on living organisms in lakes :

1. Through its **Physical properties** - medium in which organism live.
2. Through its **Chemical Properties** - source of nutrients for primary production.

Physical Properties : Specific heat, latent heat, specific gravity or density, pressure, compressibility, buoyancy, viscosity and surface tension.

Unusual Properties of Water

(A) **The heat relations of water :** Any solid or liquid when cooled, it contracts. So does water over a certain (wide) range of temperature. But the **peculiarity** of water is that it does not contract when cooled below 4°C. With further cooling it **expands** and become less dense. Expansion is gradual until the freezing point is reached; then suddenly water undergoes **drastic expansion.** Ice is only 9/10 as heavy as the same volume of water. This **Unique Density Behaviour of Water** helps in sustaining aquatic life during winter in extreme cold regions.

(B) **High specific heat capacity of water** : Water gains or loses large amount of heat without much change of water temperature. This helps in regulating earth's temperature.

(C) **Uniquely high latent heat of vapourization/latent heat of fusion** : Because of this quality of water as solid, liquid or vapour, is rivalled for storage and release of heat. Cooling produced by the evapouration of water helps animals/ plants in getting rid of excess heat.

(D) **Water is a universal solvent** : As a universal solvent, it transports through biological systems, the gases, minerals and dissolved organic components that drive life's machinery. As an inert solvent there is practically no chemical action on most of the substances it dissolves.

(E) **Boiling and melting point of water :** Unusual wide range of temperature between melting point (0°C) and boiling point (100°C) of water helps life to sustain in wide fluctuations in temperature.

(F) **Very high surface tension of water** : Excepting mercury water has the highest surface tension which is important in the maintenance of protoplasmic form and movement.

(G) Water is an exceedingly mobile liquid and is virtually noncompressible.

What makes Water Behave so Differently from other Liquid or Chemicals?

The molecular structure of water is very peculiar. Hydrogen bonding formed between the hydrogen of one water molecule and the oxygen of another adjacent molecule produce a matrix known as a liquid crystal. The maintenance of this liquid crystal gives water most of its unusual properties.

Density of Water

It is 755 times denser than air. Aquatic organisms such fishes do not have to carry their body weight in water, thus they save energy and there is reduction of their supporting tissues. Greater density exerts a marked buoyancy on aquatic organisms.

Density of water : It is maximum at 4°C. It varies :

- inversely with temperature
- directly with pressure
- directly with dissolved salt/substances
- directly with suspended particles and
- directly with buoyancy.

Hydrostatic Pressure

Water is a heavy substance (1cc weighs 1 g at 4°C). It increases with depth (1 atmosphere with 10 m depth). Objects may collapse under heavy pressure of water which is known as **IMPLOSION.** Hydrostatic pressure can lower the temperature of maximum density. The temperature of maximum density decreases about 0.1°C per 100 m depth.

Viscosity of Water

Water has low viscosity as compared to other liquids. It is 100 times greater than air.

Viscosity of water varies : **indirectly** with temperature **directly** with dissolved salts/substances.

Surface tension of Water

Water has high surface tension. It occurs due to unbalanced attraction between molecules since the surface molecules are attracted on one side only. Upward attraction is lacking since there are no water molecules above them. This acts at the air-water interface and forms a special habitat for organisms adapted to living in surface film, e.g. water striders. Surface tension of water varies indirectly with dissolved salts/substances.

Other properties of water/ice/vapour

(A) **Boiling point of water** : (100°C at 760 mm).

It **increases** with the increase of dissolved salts/substances and **decrease** with the decrease of pressure.

(B) **Freezing point of water** : (0°C at 760 mm).

It decreases with increase of dissolved salts/substances and increases with increase of pressure.

(C) Melting point of ice increases with increase of pressure.

Ice melts at 0°C, but when mixed with chemicals, melting pointis lowered. Also under additional pressure ice melts at lower temperature.

Salt : Ice.... 1.8 melts at -7°C. Seawater freezes at -2.2°C. Ice floats with 1/7 to 1/8 of its volume above seawater. Ice also flows when very thick (60m-90m thick). Specific gravity of ice at 0°C=0.9168 while sp. gravity of water at 0°C = 0.999868.

Icebergs and Glaciers

Icebergs

Icebergs are mountains of freshwater ice floating in the ocean. They are huge chunks broken off from the great masses of land-ice called glaciers. Icebergs are from about 100m to several kilometres long. Some float 3 m above water surface and may be as high as 40-120m.

Glaciers

A glacier is a mass of ice so thick that it actually moves by the force of gravity and the pressure of its own wight. They move very slowly. A few centimetre to 16 m a day (slow moving). They are of two types :

(i) Valley glaciers and

(ii) Continental glaciers.

11
Heat : Thermal Properties of Water

Heat and Temperature

The greatest source of heat to lakes is solar radiation and most of it is absorbed directly by the water. Temperature is the measurement of the **intensity of heat,** rather than the **quantity of heat.** Thermal conductivity of water is very low. Heat is distributed by the ascending and descending currents set up by **convection.**

Thermal Stratification

Sunlight falling on the lake surface heats the surface water and thus forms a layer of less dense warmer water overlying a denser cooler zone. Due to convection and wind stirring, a layer of water is formed with defined boundary. The lake is then stratified into separate thermal strata :

A. Epilimnion :

The warmer less dense upper layer is termed **Epilimnion.** This is the turbulent superficial layer of a lake lying above the metalimnion and it does not have a permanent stratification. (Fig. 6)

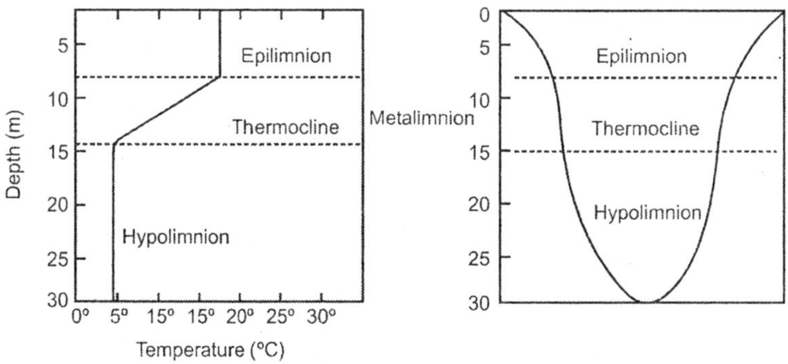

Fig. 6 : Typical thermal stratification of a lake.

B. Metalimnion : Thermocline

The stratum or layer below the Epilimnion is **Metalimnion.** The term **Termocline** is defined as the plane in which temperature falls most rapidly. The layers on either side of the thermocline, loosely embracing the whole region fall in temperature with respect to depth, are collectively called the **Metalimnion.**

(Limne - lake; clino - slope, incline; thermos - heat)

C. Hypolimnion

The deeper layer of a lake lying below the metalimnion and removed from surface influence is termed **Hypolimnion (Fig. 7).** The lack of water mixing between the warm Epilimnion and the cooler **Hypolimnion** is a major factor determining lake productivity.

Spring Overturn - Summer Thermal - Fall overturn - Winter Ice cover Stratification

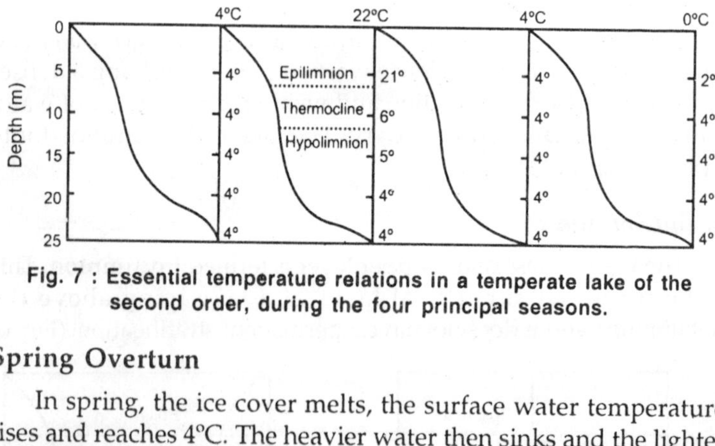

Fig. 7 : Essential temperature relations in a temperate lake of the second order, during the four principal seasons.

Spring Overturn

In spring, the ice cover melts, the surface water temperature rises and reaches 4°C. The heavier water then sinks and the lighter water rises to the surface and warms up to 4°C. This continues till the entire lake gets the same temperature from top to bottom (same density). Under the influence of spring winds, phenomenon known as **spring overturn.**

Summer Stagnation Period

When the summer thermal stratification is permanently established, the lake enters upon what is known as the summer

stagnation period since water of the hypolimnion becomes 'stagnated' for the summer and at least part of the autumn.

Fall Overturn

With the passing of summer and early autumn, air temperature declines and the surface water of lakes begins to cool down. Water thus cooled sinks and due to convection currents temperature epilimnion is lowered and finally attains the low temperature of the termocline. It gradually drops to the temperature level of hypolimnion. Wind actions then circulate the entire water from top of bottom producing the phenòmenon known as **fall overturn.**

Winter Stagnation Period

Declining air temperature in early winter lower the surface water temperature until it falls below 4°C, after which the water becomes colder and lighter. Being lighter, it no longer sinks but stays at the surface. Ultimately, the lighter surface water reaches 0°C and permanent ice cover forms on the surface and blankets off the lake completely from wind disturbances; the whole lake now resembles, in a general way, the hypolimnion o fhte summer stagnation period.

Classification of Lakes Based on Their Thermal Conditions

According to Hutchinson and Loffler (1956) and Hutchinson (1957) : (Mictic = Mixing)

Dimictic Lakes (L. Mendota, Wisconsin)

Freely circulating twice a year in spring and autumn; directly stratified in summer, inversely in winter. It represents most common type of thermal stratification, commonly found in most lakes of the cool temperate regions of the world.

Cold Monomictic Lakes (Char L. Canada)

With water temperature never excess of 4°C at any depth, freely circulating once in **summer** at or below 4°C. Largely restricted to the arctic and mountain lakes.

Warm Monomictic Lakes (L. Superior, L. Michigan)

Temperature do not fall below 4°C at any level; freely circulating in **winter**; directly stratified in summer. Such lakes are common in warm regions of temperate zones.

Oligomictic Lakes (Lakes in equatorial region)

In which water temperature is always well above 4°C; have rare circulation periods at irregular intervals; generally tropical lakes.

Polymictic Lakes (Clear L. California)

Lakes with frequent or continuous circulation (1) at little over 4°C at all depths - cold nature well over 4°C - warm polymictic troical lakes.

Amictic Lakes Antartic Lakes

No circulation, being sealed off perennially by ice from most of the annual variations in temperature.

Holomictic Lake

Those in which circulation occurs throughout the entire water column-as in the above-mentioned lakes (excepting amictic lakes).

Meromictic Lakes (Deep Lake : L. Tanganyika or Chemically stratiried lakes)

Those which do not undrego complete and the primary water mass does not mix with the lower portion. The water strata in such lakes are unique. The deeper stratum of water that is perennially stagnant is called **Monimolimnion.** This underlies the upper **Mixlimnion,** which periodically circulates. These two strata are separated by a steep salinity gradient which is called **Chemocline.** (Lakes with salts at the bottom).

Classification of Lakes : by Forel (Modified by Whipple).

I. *Polar Lakes* – Surface temperature never above 4°C.

Order 1 : Bottom water at 4°C throughout year.

Order 2 : Temperature of bottom water varies but not far from 4°C.

Order 3 : Temperature of bottom very similar to surface water.

II. *Temperate Lakes* – surface temperature varies above and below 4°C.

Order 1 : Temperature of bottom water at 4°C throughout the year.

Order 2 : Temperature of bottom varies but not far from 4°C.

Order 3 : Temperature of bottom very similar to surface water.

III. *Tropical Lakes* – surface temperature always above 4°C.

Order 1 : Temperature of bottom water near 4°C throughout year.,

Order 2 : Temperature of bottom water varies but not far from 4°C.

Order 3 : Temperature of bottom water very similar to surface water.

Head Budget of Lakes

Temperature changes in lakes are results of two sets of processes :

Addition Heat	Loss of Heat
1. Radiation from sun and sky	1. Evaporation
2. Condensation of water vapour	2. Convection to atmosphere
3. Convection from atmosphere	3. Convection to Lake basin
4. Convection from Lake Basin	4. Back radiation from lake surface to atmosphere.
5. Heat from chemical changes	
6. Heat from friction in water movements	

Gross or Crude Heat Budget

The amount of heat necessary to raise its water from 0°C to the maximum temperature found in summer.

Annual Heat Budget (Maximum Heat Content)

The amount of heat necessary to raise its water from the minimum temperature of winter to maximum summer temperature.

Summer Heat Income

The amount of heat necessary to raise its water from 4°C to the maximum summer temperature

Winter Heat Income

The amount of heat that is necessary to raise the temperature from its minimum heat content to 4°C.

12
WATER : VARIOUS MOVEMENTS

Water movements or the motion of water is of major significance on the productivity of a lake and is responsible for the distribution of all forms of energy, momentum, nutrients, dissolved gases, phytoplankton (algae) zooplankton and sedimentary material. (Fig. 8)

Wind is the main force moving lake water at all depths. The kinetic energy from the wind is transferred to the water to produce irregular turbulent.

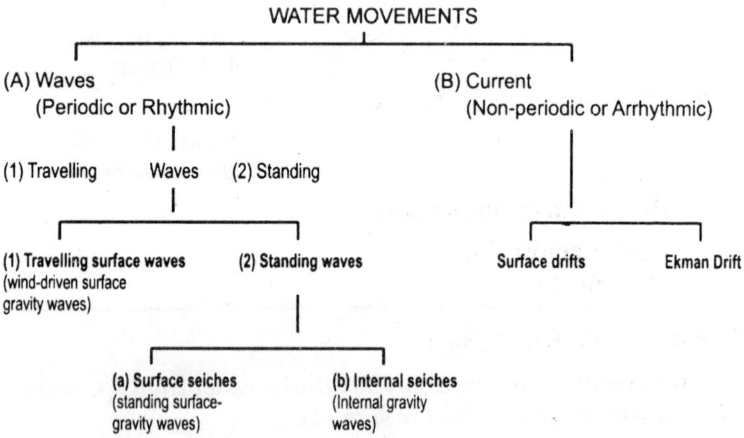

Fig. 8

A. **Waves** : Consist of rise and fall of water particles involving some oscillations **but no net flow.**

(1) *Travelling surface waves* - these are normal waves, limited only to upper-most strata.

(2) *Standing waves* - stading waves cause much greater displacements of water particles.

Seiche : Seiche is an example of standing wave and consists of a local periodic rise and fall of the water level.

Seiches are commonly generated when winds blow fairly constantly from one direction.

(a) *Surface Seicies* – These are regular, resonating waves which pass from one edge to the other and back. They occur in the boundary between air and water only.

(b) *Internal Seiches* – These have large amplitude but short wavelength and occur in the thermocline. Wind induces internal seiches.

B. **Currents** : Current are non-periodic and consist of net unidirectional flows of water

(1) *Surface drifts* – Water currents at lake surface are called surface drits. Due to earth's rotation, water currents flow at 45°C (approx.) to the direction of prevailing wind (Coriolis effect).

Coriolis effect (Coriolis force) – The diverting force of the earth's rotation which causes horizontally moving water to be diverted towards the right in the Northern Hemisphere and towards the left in Southern Hemisphere.

(2) *Ekman drift* – Water currents below the surface move at progressively greater angles to the wind with increasing depth; finally the deepest currents flow opposite to the direction of the wind. Such spiral-staircase like flow is called **Ekman Spiral.**

Lake currents range from 0 –> 30 cm

Currents have two types of flow :

(i) *Laminar flow* – It is smooth slipping of water particles around each other or an obstruction and has little drag on moving objects.

(ii) *Turbulent flow* – It is the random, chaotic tumbling of the water particles around each other or any object passing throughout the water. It causes diffusion and spreading of a patch.

Eddies : The tumbling motions of turbulent flow are termed Eddies.

Shear Stress : When wind flows at the surface of a lake, friction develops, resulting in a **shear stress** at the wind-water interface.

Shear instability or mixing occurs when the velocity difference produced by the turbulence flow across a density interface (epic & hypolimnion) is exceeded and the vortices formed.

Vertical Mixing of Water :

1. **Breaking waves** contribute vertical mixing.
2. Evaporative cooling at the surface and
3. **Convection current** are also the major vertical mixing forces.
4. In large exposed lakes, **upwelling** and **downwelling** produced by storms are a major mixing force.

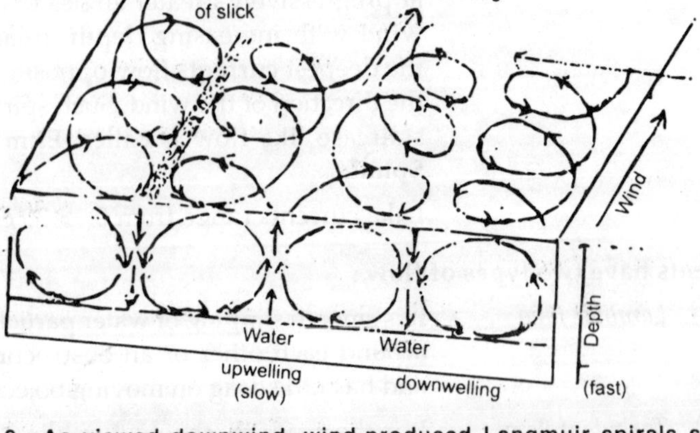

Fig. 9 : As viewed downwind, wind-produced Langmuir spirals show surface water mixed down between two adjacent left and right handed spirals.

Langmuir Spirals

These are produced due to an interaction between surface waves and wind-driven drift currents. These are the series adjacent vertical clockwise or counter clockwise rotating cells of water. They produce alternate areas of upwelling and downwelling, and foam accumulated above the downwelling zone, known as slicks or wind. (Fig. 10)

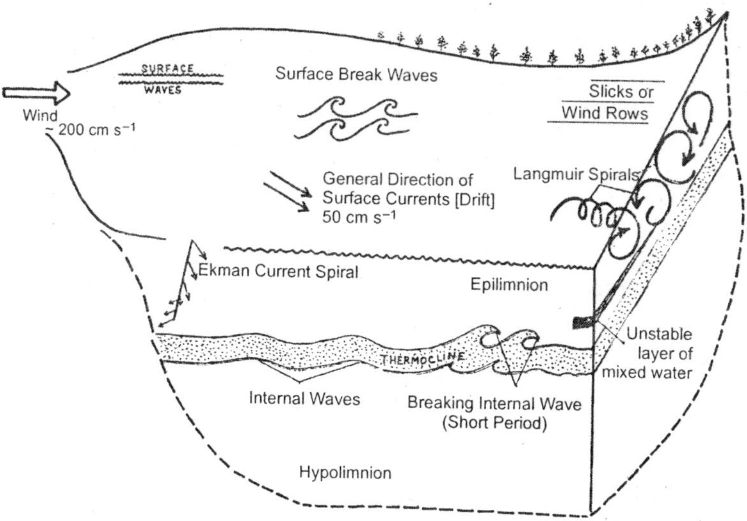

Fig. 10 : Composite diagram of forces (wind, gravity, evaporation, Earth's rotation) and the resultant water current and waves. Wind moves the water, gravity makes horizontal flow easier than vertical, evaporation cools surface water, which then sinks, and the Earth's rotation moves surface flows to the right (Northern Hemisphere) and to the left (Southern Hemisphere).

13

CHEMICAL PROPERTIES OF WATER

Water contains a vast array of inorganic and organic compounds which are present as dissolved solids and gases. These come from the soil, air and from metabolism of organisms.

The climate, watershed (drainage), atmosphere (rain water to small extent) also influence the chemical composition of lake waters.

Inorganic nutrients provide the chemical constituents upon which the entire food chain is based. There are three groups :

1. **Major and minor nutrients :** These limit growth of plants and animals in a lake. Major nutrients needed in large quantities are CO_2, O_2, NH_4, NO_3, PO_4, NO_3, PO_4, SO_4, SiO_2 and Fe. Important minor nutrients (trace elements) are Mn, Co, Mo, Cu and Zn.
2. **Those whose effect is largely ionic :** Na, K and Cl ions are often abundant in lakes but their effect is largely ioic.
3. **Toxicants :** Some common inorganic substances such as Cu and Zn act either as toxicants or as growth stimulators.

Small quantities or organic compounds are common in natural waters. Animals and also plants excrete organic compounds. Bacteria, fungi, and some protozoans grow on decomposing organic compounds.

DISSOLVED GASES

Effect of pressure and temperature on the solubility of gases.

HENRY'S LAW

At constant temperature, the solubility of a gas in a liquid is directly proportional to the pressure at which the gas is supplied.

Conditions affecting solubility of a gas :

1. Solubility decreases with rise of temperature.
2. Solubility decreases with increasing concentration of dissolved salts.
3. Rate of solution is greater when the gases are dry (free from water vapour).
4. Rate of solution depends on the degree of under-saturation of the water with the gases concerned. The greater the degree of under saturation the greater the rate of solution.
5. Rate of solution is increases by wave action and other forms of surface-water agitation.

Solubility of Some Gases in 1 Litre of Water

Name of gas	*Volume in (L) Absorbed At 0°C & 760 MM*	*Volume in (L) Absorbed At 20°C & 760 MM*
Ammonia (NH_3)	1300°	710°
Hydrogen Chloride (HCl)	506°	442°
Carbon dioxide (CO_2)	1.713°	0.878°
Oxygen (O_2)	0.049°	0.031°
Nitrogen (N_2)	0.0239°	0.0164°
Hydrogen (H_2)	0.0215°	0.0184°

Dalton's Law of Partial Pressures

The pressure exercised by each component in a gaseous mixture is proportional to its concentration in the mixture, and the total pressure of the gas is equal to the sum of those of its components.

Volume of a gas varies inversely with the pressure, but the actual volume of a gas absorbed in a liquid (water) varies direcly with pressure of the dissolved gases in natural waters. The following are very important :

O_2, CO_2, CH_4, H_2S, N_2, NH_3, SO_2 and CO.

Dissolved Solids

In normal fresh waters, the dissolved solids consists largely of only a few salts. The common salts are - carbonates ($-CO_3$), sulfates ($-SO_4$), and chlorides of calcium, magnesium, sodium and potassium, silicic acid and small amounts of nitrogen and phosphorous compounds. In addition, there are compounds of iron and manganese (seldom). There are also some dissolved organic substances.

Quantity of dissolved solid present in lake water can be determined by evaporating one litre of water and weighing the residue.

When a gas comes in contact with water, it dissolves in the water until a state of equilibrium is reached in which the solution and the emission of the gas are balanced.

As for example, -oxygen (gas) $\rightleftarrows$ oxygen (dissolved)

Salinity of Inland Waters

Salinity of inland surface waters has an average concentration of 120 mg^{-1} (world average) but varies appreciably depending on its surface.

Ionic Composition of Freshwaters

The concentration of four major cations Ca^{2+}, Mg^{2+}, Na^{+} and K^{+} and four major anions $HCO\text{-}_3$, $CO^{2-}{}_3$, $SO^{2-}{}_4$, and Cl^- usually constitute the **total ionic salinity** of water for all practical purposes. Those of Si (in hard waters), N, P and Fe though of immense biological importance, are very small.

Major Ions in Freshwaters and their importance

(a) *Calcium*

Ca is important for the growth of freshwater flora and fauna. It is required as a nutrient of normal metabolism of higher plants. It is also considered a basic inorganic element of algae. Ca is essential for maintenance of the structural and functional integrity of cell membranes in ion absorption and retention. The Ca content of softwater lakes remains well below saturation levels. In hardwater lakes, Ca concentration however, undergoes marked seasonal dynamics.

(b) *Magnesium*

Mg is required by all chlorophyllous plants as the Mg-porphyrin component of the chlorophyll molecules and as a micro-nutrient in enzymatic transformation of organisms. Low available Mg has been observed as one of the several factors influencing phytoplankton productivity in extremely oligotrphic lakes.

(c) ***Sodium and Potassium***

These are primarily involved in ions transport and exchange. Na requirement is particularly high in some blue-green algae. A threshold level of $4mgl^{-1}$ is required for near optimal growth of several algal species. Moderate epilimnetic reduction in K has been observed in extremely productive lakes, presumably due to the utilization of K by the legal population.

(d) ***Chloride***

Cl^- is not dominant in open lake system. Lakes situated near the sea receive significant inputs of chlorides from atmospheric transport from the sea. **Chloride** is influential in general osmotic salinity balance and ion exchange.

Sources of Salinity :

A. *Weathering of Soil and Rock :*

The composition of soil and rock and their exchange capacity influence the rates of weathering and ion supply to the percolated and run off waters. Oxidation and reduction processes primarily affect Fe, Mn, S, N, P and C compounds in soils. Oxidation of sulphides are the major sources of sulfate in natural waters. The action H^+ ions from the dissociation of carbonic acid is of major importance in the weathering of rocks and soils.

The salinity of natural waters is further influenced by the depth and mode of water percolation.

B. *Atmospheric precipitation and fall out :*

Rainfall carries much of the atmospheric salts to lakes and rivers. Continental rain generally contains more sulfates than chlorides. The chloride ions concentration is more abundant in lakes near the sea.

Wind-blown dusts from the soil often contribute salts especially of Ca and K to rain and snow. An additional major source of atmospheric salinity is industrial and domestic air pollution. Of the major ions emitted into the air, **chlorides,** Ca and SO_4 are major contaminants.

C. *Environmental Influence on Salinity*

Climate has a marked effect on the balance between precipitation and evaporation and on the salinity conc. of surface waters. Evaporation increases the **Salinity** of closed drainage basins. Other climatic factors are **temperature** and **wind**. Temperature influences the rate of rock weathering. Wind speed and direction affect the chemical composition of atmospheric precipitation.

OXYGEN : The Oxygen Content of Freshwaters

Dissolved oxygen (D.O.) is essential for the metabolism of vast majority of the aquatic organisms.

Sources of Dissolved Oxygen in Natual Waters

(1) Directly from the atmosphere through the exposed surface. Air contains by volume about 20.95% oxygen. Absorption of oxygen from air :

(a) by direct diffusion at the surface.

(b) through various forms of surface agitation e.g. wave action, etc.

The Principal Causes of O_2 Decreases in Water :

1. Respiration of animals and plants.
2. Decomposition of organic water.
3. Reduction due to other gases.
4. Automatic release of D.O. from epilimnetic water due to summer temperature.
5. Inflow of sub-terranean water.
6. Presence of iron (Fe).

Solubility of Oxygen

Solubility of O_2 is reduced both by temperature and salinity. O_2 is about 1/5 less soluble in seawater than fresh water.

Table 3 : Solubility of O_2 in pure water in relation to temperature from saturated air at 760mm/ pressure

Temperature	*O_2 mg/l of ppm*	*Temperature*	*O_2 mg/l or ppm*
0°C	14.16	20°C	8.84
5°C	12.37	25°C	8.11
10°C	10.92	30°C	7.53
15°C	9.76	35°C	7.04

Oxygen Profile in Lakes (Types of O_2 – Depth Curve)

Four general types of O_2 distribution in lakes are recognized.

(a) *Orthograde curve*

In oligotrophic lakes (less productive or unproductive lakes), O_2 distribution is largely a function of temperature resulting in fairly uniform Orthograde curve during summer stratification. The O_2 concentration in circulating epilimnion decreases with rise of water temperature. But in the meta and hypolimnion, the O_2 concentration increases because of decreasing temperature and remains at saturation at the hypolimnion from just before the summer stratification.

(b) *Clinograde Curve*

In very productive **Eutrophic lakes,** O_2 depletion occurs in the hypolmnion during summer stratification and also during winter stagnation under ice cover due to oxidative processes. The resulting curve in the hypolimnion is anaerobic and is termed a **Clinograde curve.** This is characterized by relatively higher O_2 content near the surface where photosynthesis replenishes the supply.

(c) *Heterograde Curve*

It exhibits and irregular slope from the lake surface to the depth. It is of two types :

(1) *Positive Heterograde Curve or Metalimnetic O_2 Maxima* : In this, there is an increase of O_2 in the metalimnion during stratification. As the solubility of O_2 decreases in the epilimnion with increasing summer temperature and O_2 consumption in hypolimnion increases, metalimnetic O_2 maxima may occur due to photosynthetic activity of phytoplankton (O_2 - producers) which concentrates at the region.

(2) *Negative Heterograde Curve or Metalimnetic O_2 Minima* : This is the converse condition. A metalimnetic O_2 minima exhibiting a negative heterograde – curve is usually caused by concentrations of animals when, O_2 consumption due to respiration and decomposition dominantes at some mid-depth.

(d) *Anomalous Curve*

This may result from the settling of cooled, highly oxygenated surface water or by layering at intermediate depths of inflow waters that have a different O_2 concentration than the lake in questions.

Oxygen Deficit

Oxygen deficit in a lake is equal to the amount of O_2 necessary to reach saturation **minus** the amount of O_2 actually present.

Diel Changes in O_2 Content

Diel means 24 hours : day and night.

Nocturnal means night only and **Diurnal** means only the lighted period of the day.

Measurements of the distribution of O_2 in lakes is very important for limnological studies. The concentration of O_2 in a lake is function of biological processes such as, photosynthesis or respiration and other physical processes such as water movement or temperature.

A highly productive lake may be below saturation to supersaturation in the course of 24 hours. Even the O_2 may get nearly depleted during early morning hours.

Action of O_2 on Iron

Oxygen combines readily with other elements to form oxides. If sufficient soluble Fe is present then Ferric oxide is formed. The change in oxidation state of metal ions and some nutrient compound is defined by the **Redox** or **Oxidation-reduction potential.**

Carbon Dioxide (CO_2) : Carbon Dioxide Contents of Freshwaters

Carbon is the primary element in organic syntheiss. CO_2 provides the major source of carbon in photosynthesis and is a product respiration by both plants and animals.

Sources of CO_2 in Water

1. By direct diffusion from air. CO_2 is about 0.033% by volume in air.

2. From respiration of animals and plants.
3. By decomposition of organic matter.
4. From inflowing groundwater.
5. In combination with other substances mainly Ca and Mg reduction of CO_2 in natural waters.

Reduction of CO_2 in Natural Waters

1. Photosynthesis of aquatic plants.
2. Agitation of water.
3. Marl-forming organisms.
4. Evaporation.
5. Rise of bubbles from depths.

Solubility of CO_2

CO_2 dissolves readily in water. The soubility of CO_2 is about 200 times that of O_2. It decreases with increase in temperatures. Solubility of CO_2 increases considerably in water that contains carbonates.

Table 4 : Amount of CO_2 Dissolved in water from air at sea level

Temperature	CO_2 mg/l
0ºC	1.1
15ºC	0.6
30ºC	0.4

CO_2 also occurs in nature in other important forms.

(a) Bound CO_2 - as nearly insoluble monocarbonate such as $CaCO_3$ and $MgCO_3$

(b) Half-bound CO_2 - that additional amount required to convert $CaCO_3$ or $MgCO_3$ into the bicarbonate, such as, $Ca(HCO_3)_2$

CO_2 dissolves in water to produce carbonic acid (H_2CO_3) which fairly weak acid and dissociates into various fractions depending upon the hydrogen-ions concentration (pH).

$$CO_2 + H_2O \rightleftharpoons CO_2 + H_2O \quad (A)$$
$$H_2CO_3 \rightleftharpoons H^+ + HCO_3^- \quad (B)$$
$$HCO_3^- \rightleftharpoons H^+ + CO^{2-}_3 \quad (C)$$

[$CO_2 + H_2CO_3 + CO^{2-}{}_3$] is known as the **total inorganic carbon or Σ CO_2]**

What is pH?

Pure water dissociates weakly H^+ and OH^- ions. The dissociation constant is very small (10^{-14}). The amount of H^+ and OH^- present are 10^{-7} g ions per litre.

pH is defined as :

(1) Logarithm of the reciprocal of the concentration of free hydrogen ions [H^+]

(2) It is the negative logarithm of the Hydrogen ion concentration (H^+) expressed in gram equivalents.

In neutral water $H^+ = OH^-$ (10^{-7}) which means pH is 7; when more H^+ ions are added (due to acid reaction). The power neutrality (10^{-7}) is increased to say 10^{-5}, i.e. pH is 5 which is acidic. Where the reaction more alkaline H^+ decrease, say to 10^{-9}, i.e. the pH is H^+ decrease, say to 10^{-9}, i.e. the pH is 9 and the water is alkaline.

Acidity and alkalinity of water

Acidity of natural waters is due to total available acid and the concentration of hydrogen gas.

Alkalinity of natural waters

This refers to the quantity and kinds of compounds present in water which collectively shifted pH to the alkaline side. In freshwaters, alkalinity is mainly due to presence of bicarbonates, carbonates and hydroxides.

Total Alkalinity

The milliequivalents of acide necessary to neutralize the HCO^-_3, CO^{2-}_3 and OH^- ions in one litre of water determines the total alkalinity.

Free CO_2 dominates in water pH 5 and below HCO^-_3, predominates between pH 7 and 9 while CO^{2-}_3 is quantitatively significant above pH 9.5.

The Carbonate Buffer System : What is a Buffer?

Mixtures of weak acides and their salts are called **Buffers.** In natural waters, carbonates are the most significant buffer system and is primarily responsible for preventing rapid changes of pH in aquatic ecosystem.

Weak acid H_2CO_3 (carbonic acid). Its salts : $CaCO_3$ and $MgCO_3$ and Ca $(HCO_3)_2$. As we know from equations (A), (B) and (C) the ions present are H^+ and HCO^-_3 and CO^{2-}_3.

(1) Hence, if H^+ ions (or an acid) is added to the water is reacts with HCO^-_3 to form CO^{2-}_3 and are removed from the solution thus preventing a decrease in pH or making the water acidic.

$$H^+ + HCO^-_3 \rightleftharpoons H_2CO_3\ H_2O + CO_2 \qquad [D]$$

(2) If OH^- is added, then it reacts with HCO_3^{-1} to form CO^{2-}_3 thus preventing the water to become alkaline.

$$OH^- + HCO^-_3 \rightleftharpoons CO^{2-}_3 + H_2O \qquad [E]$$

(3) If CO_2 is added (due to respiration and decomposition of organic matters) then it (CO_2) will combine with CO^{2-}_3 to form HCO^-_3 thus preventing drop in pH due to release of H^+

$$CO_2 + CaCO_3 + H_2O \rightleftharpoons CaCO^{2-}_3\ (HCO_3)_2 \qquad [F]$$

(4) If CO_2 is withdrawn from water (due to photosynthesis), the HCO^{-1}_3 dissociates into CO^{-2}_3 or H_2CO_3 thus preventing increase in pH.

$$HCO^-_3 \rightleftharpoons H^+ + CO^{2-}_3 \qquad [C]$$

$$Ca(HCO_3)_2 \rightleftharpoons CO_2 + H_2O + CaCO_3 \qquad [G]$$

(Reverse of F).

Classification of Lakes : On the basis of bound-CO_2

(a) *Softwater lakes* – Those waters hold small amounts of Ca and Mg salts in solution and in which the average bound-CO_2 does not exceed 5ml, l^{-1}

(b) *Medium-class lakes* – Those whose bound-CO_2 falls between 5 and 22 ml l^{-1}

(c) *Hardwater lakes* – Those which contain more than 22 ml l^{-1} bound-CO_2. The maximum may be about 50 ml, l^{-1}

Nitrogen

After C, H and O, the most abundant element in living cells is **nitrogen** which is essential for most biochemical reactions. one to 10% of N by dry weight is present in animals and plants.

Sources of N in water :

(A) Precipitation on the water surface.

(B) Nitrogen fixation in water and sediments.

(C) Inputs from surface and groundwater drainage.

Losses of N occur by :

(A) Effluent outflow from the basin.

(B) Reduction of NO_3^- to N_2 by bacterial denitrification; N_2 espace to air.

(C) Permanent loss of inorganic and organic N-containing compounds to sediments.

Forms of N in Lake Waters

Nitrogen occurs in freshwaters in numerous forms : dissolved molecular N_2, and NH^+_4, NO^-_2 and NO^-_3 and also in a large number of organic compounds (amino-acids, amines and proteins).

(a) N_2 gas is most abundant because the surface waters are in constant contact with the atmosphere [80% N_2 (approx.)]

(b) **Ammonia** : It is generated by heterotrophic bacteria from decomposition of organic matters and also a major excretory product of animals. NH_3 is mainly present in water as the dissociated ion NH^+_4 which is rapidly taken up by plants (phytoplankton and macrophytes). NH_3 is highly soluble in water but is very toxic to aquatic animals.

$$NH_3 + H_2O \rightleftharpoons NH_4OH \rightleftharpoons NH_4^+ + OH^-$$

Undissociated NH_4OH is also toxic, but NH^+_4 is harmless, although NH_3 is a good source of N for plants, most plants grow better with NO^-_3 as the source of N.

Ammonification

Most of the N in organic matter exists as amino-acids, amines and proteins. Through microbial activity (bacteria) proteins, amines and amino-acids are deaminated and NH_3-N is produced. This process is called **ammonification and** $\underline{NH_3}$ is released to environment or assimilated into microbal tissue.

Nitrate (NO^-_3) and Nitrite (NO^-_2)

Nitrate is the most abundant form of combined inorganic **nitrogen** in lakes and streams. Nitrate is assimilated by aquatic plants being reduced to **ammonia.** Nitrite, the partially reduced form of nitrate is usually present in insignificant quantities. Nitrates are mainly formed by oxidation from nitrogenous compounds **(nitrification).**

Nitrification

It is broadly defined as the biological conversion of organic and inorganic nitrogenous compounds from a reduced state to a more oxidized state. The nitrifying bacteria **ntirosomonas** oxidizes NH^+_4 to NO^-_2 and obtain energy.

$$NH^+_4 + 1\tfrac{1}{2}O_2 \rightleftharpoons 2H^+ + NO^-_2 + H_2O$$

Oxidation of nitrite proceeds further to nitrate by the bacteria **Nitrobacter**

$$NO^-_2 + \tfrac{1}{2}O_2 \rightleftharpoons NO^-_3$$

The overall nitrification reaction is :

$$NH^+_4 + 2O_2 \rightleftharpoons NO^-_3 + H_2O + 2H^+$$

The nitrification is most rapid at pH 7-8 and at temperature 25-35°C.

Denitrification

In the absence of O_2 many micro-organisms can use NO^-_3 and NO^-_2 instead of O_2 in respiration. The heterotrophic process is termed **denitrification** because N_2 (gas) is released as metabolite and is lost from the system.

Reduction of nitrate (1)

$$C_6H_{12}O_6 + 12NO^-_3 \rightleftharpoons 12NO^-_2 + 6CO_2 + 6H_2O$$

Reduction of nitrate (2)

$$C_6H_{12}O_6 + 8NO^-_2 \rightleftharpoons 4NO_2 + 2CO_2 + 12NO^{2-}_3 + 6H_2O$$

The following facultative anaerobic bacteria are able to denitrify nitrate to N_2 gas :

Pseudomonas, Achromobacter, Escherichia, Bacillus and **Micrococcus**

Nitrogen Fixation

N_2 gas although abundant in water is almost inert and is normally utilized for growth in lakes and streams by some blue-green algae and bacteria through nitrogen fixation.

(A) **By blue-green algae :** Nitrogen is fixed in waters mainly by the heterocyst-containing blue-green filamentous algae such as, **Anabaena, Nostoc, Gleotrichia, Nodularia** and others. **Anabaena** associated with the floating weed **azolla** also fixes N_2.

(B) **By Bacteria :** The common heterotrophic N_2-fixing bacteria are several species of azotobacter and clostridium and by photosynthetic bacteria.

(c) By the dense stands of non-leguminous angiosperm of genus **Alnus** and **Myrica**

Nitrogen fixation is positively correlated with concentrations of dissolved organic nitrogen in water. Nitrogen-fixing algae and some photosynthetic bacteria can fix only limited quantities of N_2 in the dark. The enzyme responsible for nitrogen fixation is called **Nitrogenese.** Under anoxic conditions found in the hypolimnion of some lakes, photosynthetic N_2-fixing bacteria may important.

The rate of **denitrification** and also of **nitrification** decreases in acid waters and is very slow at low **temperatures.** The denitrification reactions occur intensely in anaerobic environments, such as in the hypolimnia of eutrophic lakes.

Seasonal Cycles of Nitrate

The major changes occurring between autumn and winter and between winter and spring are considered in three categories.

1. *Oligotrophic lakes* in which nitrate levels remain almost constant (L. Superior).
2. *Eutrophic lakes* in which NO^-_3 levels fall to almost zero (L. Clear).
3. *Mesotrophic lakes* in which NO^-_3 levels fall but may not become limiting (L. Mendota).

Phosphorus (P)

Phosphorus is least abudnant in nature and yet it limits the biological productivity of natural waters. Free P does not occur in nature.

Sources of Phosphorus in inland waters :

(a) Precipitation.
(b) Land run off and inflowing waters.
(c) Groundwater.
(d) Plant debris and animal activity.

Loss of Phosphorus

(a) Outflowing river.
(b) Bottom deposits.
(c) Plant and animal harvest. Phosphorus is removed from water phytoplankton.

Forms of Phosphorus in natural waters

(A) *Inorganic Fractions:*
  (1) Solubble phosphates
  (2) Sestonic acid soluble phosphate - P
      (Mainly $FePO_4$ and $Ca_3 (PO_4))_2$
(B) *Organic Fractions:*
  (1) Dissolved and colloidal P (30%)
  (2) Organic Sestonic P (70%)

The total P in lakes varies from less than 1 mg l^{-1} to even 78 mg l^{-1}. It increases with lake productivity. Lakes reach in organic matter have higher total P concentrations.

Oligotrophic lakes show little variaton in P content with depth, whereas Eutrophic lakes show a considerable increase in the P content of the lower hypolimnion. This is primarily due to liberation of phosphates from sediments. In oligotrophic lakes with orthograde curve, oxidized metals such as Fe and major cations particularly Ca induce precipitation of Phosphorus.

Soluble Inorganic Phosphates

Soluble orthophosphate ions are the simplest form of Phosphorus in water.

$$H_3PO_4 \rightleftharpoons H^+ + H_2PO^-_4$$

$$H_2PO_4 \rightleftharpoons H^+ + HPO^{2-}_4$$

$$HPO^-_4 \rightleftharpoons H^+ + HPO^{3-}_4 \quad \text{(Orthophosphate ion)}$$

Presence of ionic species depends of pH.

Soluble orthophosphate concentrations are usually not more than 5-20 mg l^{-1} and P and seldom exceeds 0.1 mg l^{-1} even in highly eutrophic waters. The water in oligotrophic lakes often contains less than 0.001 mg of inorganic P per litre.

Phosphorus Cycling

Lake sediments contain much higher concentration of P than the water. The exchange of P between the sediments and overlying water is a major component of the P-cycle in natural waters. There is a net movement of P into the sediments especially of genera **Pseudomonas, Basterium** and **Chromobacterium** and abundant in lake sediments. The rate of release from the sediments increases markedly, if the sediments are distrubed.

Phosphorus is essential for plant growth although needed in a very small quantity as compared to C and N. The C:N:P ratio of plant roughly 40C:7N:1P by weight.

Iron (Fe)

Apart from C, H, O, N and P, iron (trace element) is needed by most living organisms in larger quantities. Fe is essential for many enzymatic and other cellular processess. Fe exists in water in solution as either Ferrous (Fe^{++}) or Ferric (Fe^{+++}) state. Amount of Fe in solution and the rate of oxidation of Fe^{++} in oxic (aerobic) water depends on **pH, Eh and temperature.**

Redox Potential

Eh is the redox potential which is an expression of the oxidizing and reducing power of a solution. This power is dependent on the proportion of its oxidized or reduced components in the solution. Redox potential is greatly influenced by pH.

Fe^{++} is more soluble than Fe^{+++}. Soluble Fe^{++} occurs mainly as $Fe\ (OH)_2$, but is insoluble at normal pH is oxygenated natural waters. The most common species of Ferric ion in natural water is $Fe\ (OH)_3$. It's solubility is very low.

Vertical distribution of Fe in lakes

The vertical distribution of the Fe is reflected in the distribution of redox potentials. Ionic Fe of oxygenated waters of oligotrophic lakes an epilimnia of eutrophic lakes is very low. In the hypolimnion of eutrophic lakes however, when decomposition continues during the summer stratification period, the Eh falls below 100 mv and H_2S is produced which **precipitates free F^{++} to FeS.**

Manganese, Calcium and Magnesium

Manganese

Mn is also micronutrient (trace element) whose cycling natural waters closely parallels Fe. It is needed by plants for the O_2 evolving process in photosynthesis and also in several enzyme systems as a co-factor.

Calcium

Ca is essential for metabolic process in all living organism and as a structural or skeletal material in many. Ca has an important role in its effect on pH and CO_2 - HCO^-_3 system. Ca is present in ionic form. Carbonates and bicarbonates of both Ca and Mg constitute the main buffering system in natural waters.

Magnesium

Mg is often associated with Ca primarily due to its similar chemistry. It is present in large amounts in aquatic ecosystem. Mg is needed by all cells for phosphate transfer involving ATP and ADP.

$$ATP^{Mg2+} \rightleftharpoons ADP + P + Energy$$

Sulphur (S)

Sulphur in the form of both mineral and organic sulphates is utilized by all living organisms. Sulphate is needed in protein synthesis, in photosynthesis and in animal metabolism. Several groups of bacteria oxidize S to sulphide and sulphide to sulfate. Other groups reduce sulfate to H_2S.

Sulphur oxidizing bacteria : They are differentiated into two groups.

(A) **Chemosynthetic sulphur-oxydizing bacteria (colourless):**

They are mostly aerobic forms that oxidize H_2S. They are of two types :

(i) The first type deposits S inside the cell

$H_2S + ½O_2 \rightarrow S + H_2S$ which accumulate as long as H_2S is available. As sulphide is depleted, the internally stored S is oxidized to sulphate :

$$S + 1½O_2 + H_2O \rightarrow H_2SO_4$$

(ii) The second type deposits S outside the cell.

$$2Na_2S_2O_3 + O_2 \rightarrow 2S + Na_2SO_4$$

(B) **Photosynthetic sulphur-oxidizing bacteria (coloured) :**

They are anaerobic forms and need light.

$$CO_2 + 2H_2S \rightarrow (CH_2O) + H_2O + 2S$$

$$2CO_2 + 2H_2O + H_2S \rightarrow 2(CH_2O) + H_2SO_4$$

Sulphur-reducing bacteria

They are strictly anaerobic and derive O_2 from SO_4 for oxidation of either organic matter or H_2.

$H_2SO_4 = 2(CH_2O) \rightarrow 2CO_2 + H_2O + H_2S$ or

$H_2SO_4 + 4H_2 \rightarrow 4H_2O + H_2S$

In stratified oligotrophic (unproductive) lakes under oxic condition H_2S is rapidly oxidized and is totally absent, while SO_4 concentrations change little with depth. In the hypolimnia of

Eutorphic (productive) lakes reduction of SO_4 to H_2S occurs due decomposition as the redox potential goes down to below 100 mv. H_2S near the sediments reacts with Fe^{++} to form insoluble FeS.

Silicon (Si)

Silicon does not occur nature as free element. Silica (SiO_2) occurs in relative abundance in natural waters in the form of :

(A) Dissolved silicic acid (at normal lake pH values)

(B) Particulate silica.

The silica content of drainage to natural waters is about 13 mg of SiO_2 per litre. silica exists in many forms but only silicic acid (H_2SiO_4) can be used by algae. Silica can be a limiting factor for phytoplankton growth where diatoms are the predominant algae.

Diatoms assimilate large quantities of silicon in the synthesis of their **Frustule** cell structure. Frustules, which are rigid cell-walls of diatoms account for half the dry weight of cells. Below a concentration of 0.5 mg of SiO_2 most diatoms cannot grow well. Some chrysophycean flagellates and a few other algae possess silicon cell-walls. Also, spicules of freshwater sponges are formed of silicon. Silica in diatoms is insoluble and less easily recycled. Some diatoms release up to 15% of the silica they take up.

During summer stratification in the anoxic (anaerobic) deep waters of lakes having clinograde O_2 curves, the silica conc. is usually increased. This is regulated by redox potential, probably by the reduction of Ferric silicate formed in mud in contact with water. Uptake of dissolved silica by diatoms involves an aerobic process.

Some common lake diatoms are : **Cyclotella, Melosira, Asterionella, Tabellaria, Navicula** and **Fragellaria.**

Littoral Zone

The shoreward region of a body of water (lake/ocean) is known as the littoral zone. It extends from the region of highest water level to a depth where light is barely sufficient for rooted aquatic plants to grow.

Pelagic/Limnetic zone (Region of the **free water** in lakes)

In deep lakes, the area beyond the influence of the shore or bottom is termed pelagic or limnetic zone.

Trophogenic or photic zone : The zone extends from the lake surface to where light is about 1% of that at the surface. Photosynthetic production occurs at this zone.

Tropholytic or Aphotic Zone : This zone extends below the littoral or photic zone to the bottom of the lake. Here light level is too low for photosynthesis. This is a region of O_2 consumption and heterotrophic decomposition of organic matter.

Profoundal zone : The deep zone of lakes below the light-controlled limit of plant growth.

Compensation point/Compensation Depth : The light intensity at which O_2 production by photosynthesis and O_2 **consumption** by respiration of the plants concerned are equal, is termed as **compensation point** and the depth at which it occurs is called the **compensation depth.**

14

CLASSIFICATION OF ORGANISMS IN WATER

I. **Seston** : Entire heterogenous mixture of living and non-living bodies which float or swim in water (sistos : strainable).

A. **Bioseston** - living components of seston

1. **Plankton** : organisms to relatively small size, mostly microscopic having low power of locomotion; drift in water, subject to the action of waves, currents and other forms of water motion. (Planktons : wandering)

2. **Neuston** : organisms related to surface film of water (Neust : swim)

   (a) Infraneuston : hanging from or floating against lower side e.g. cladocerans, pond snail

   (b) Supraneuston : hanging upon surface film, e.g. water striders, water mesurers.

3. **Pleuston** : higher plants which float either upon the surface or within the water (pleust: float/sail)

4. Nekton : organisms or larger size which swim freely regardless or movements of surroundings water (nextos : swim)

B. **Abioseston** : non-living component of seston.

1. Autochtonuous detritus : detritus produced within a given locality (auto : self; chthon : land)

2. Allochthonous detritus : detritus originating outside a given locality (allos : other).

II. **Aquatic biota of long standing :** (They are grouped as–)

A.	**Littoral**	: organisms inhabiting region adjacent to the shore. (Littor : seashore)
B.	**Limnetic**	: organisms inhabiting the open waters.
1.	True limnetic	: organisms thriving in open waters only.
2.	Limnetic	: organisms which spend part of their life cycle in open waters.
C.	**Profoundal**	: organisms living in deep region below the light-controlled limit of plant growth (profound : deep)
D.	**Benthos**	: organisms which live in or on the bottom of the basin (benthos : depth)

*Authochthonous detritus : Detritus produced by plankton, neuston, pleuston, nekton and benthic organisms.

15

Classification and Technology of Plankton

I.	**On the basis of quality**	: or (type of carbon used as food)
	Phytoplankton	: plankton of plant origin e.g. **Chlorella, Euglena**
	Zooplankton	: plankton of animal origin e.g. **Brachionus, Cyclops, Moina**
II.	**On the basis of size**	: (1mm = 1000 microns (um) or 1 micron (µm) = 0.001mm)
	– Megaloplankton	: Largest size plankton > 1 cm (Megalo : very large)
	– Macroplankton	: Large plankton; visible to naked eye 1mm - 1 cm (Macro : big)
	– Mesoplankton	: size between macro-micro-plankton 500 µm 1 mm (meso : medium)
	– Microplankton	: small-size plankton (net plankton 50 µm - 500 µm (micro : small)
	– Nannoplankton	: pass through net-mesh size 50-60 um 5-50 um (nanno : dwarf)
	– Ultraplankton	: minutes plankton : observed by electron microscope<5µm (ultra : beyond)
III.	**On the basis of environmental distribution :**	
	– Limnoplankton	: Lake plankton (Limnos : lake)

– Rheoplankton	:	Running-water plankton (Rheo : flow, current)
– Heleoplankton	:	Pond or marsh plankton (Heleo : marsh)
– Haliplankton	:	Salt-water plankton (Halo : salt)

IV. **On the basis of origin :**

– Autogenetic plankton	:	Plankton produced locally
– Allogenetic plankton	:	Plankton introduced from other locality.
– Tychoplankton	:	Forms of the littoral community occurring in plankton accidentally.

V. **On the basis of content :**

– Euplankton	:	True plankton (No debris) (EU : good, true)
– Pseudoplankton	:	Debris mingled in plankton (Pseudo : False)

VI. **On the basis of life history :**

– Holoplankton	:	Organisms free-floating throughout their life (Holo : whole)
– Meroplankton	:	Organisms free-floating only at certain times or stages of life history (Mero : part)

Please note : 'Plankton' is used both for singular and plural. Single organism in known as **Plankter.**

16

Functional Classification of Organisms

Organisms can be classified at the functional level by their sources of energy, sources of carbon and in case of lower organisms such as sulfur bacteria by the molecule which serves as the electron donor.

Classification as :

I. *Source of energy*

A. **Phototrophs** : organisms which derive energy directly from sunlight through photosynthesis.

B. **Chemotrophs** : organisms which utilize chemical energy as the source.

II. *Source of carbon*

A. **Autotrophs** : organisms which utilize inorganic CO_2 to produce organic matter.

B. **Heterotrophs** : organisms which depend on performed organic carbon such as glucose.

III. *Source of electron*

A. **Organotrophs** : the cells of organisms receive electrons from organic compounds

B. **Lithotrophs** : organisms that derive electrons from inorganic matter.

Algae are differentiated as :

Auxotrophs : organisms (algae) that require vitamins for growth

Phto-autotrophs : they require light energy for cell metabolism

Heterotrophs : they sustain growth and cell division in the dark; energy and cell carbon are obtained from metabolism of an organic substrate.

Mixtrophs : they assimilate CO_2 in small amounts simultaneously with organic compounds both in light and (especially) in dark.

Cyanophyta/Myxophyta (Blue-green algae)

General Characteristics :

1. Photosynthetic, prokaryotic micro-organisms; some capable of N_2-fixation
2. Mitochondria, golgi apparatus, plastids and chloroplasts absent; also lack membrane-bounded DNA.
3. Spores resisted to adverse conditions.
4. Heterocysts : thick-walled translucent; formed of vegetative cells; site of N_2-fixation.
5. Unicellular, colonial or filamentous forms; all the forms having gelatinous sheath.
6. Reproduction primarily by binary fission.

Example : **Anabaena, Nostoc, Oscillatora, Gleotrichia, Microcystis (Fig. 11)**

Chlorophyta (Green-algae) :

General Characteristics :

1. Unicellular, colonial or multicellular forms
2. Having chlorophyll a and b.
3. Cell-wall of cellulose or pectin
4. Reverse food - starch.

Example : **Chlorella, Scenedesmus, Volvox, Pediastrum** and **Desmids** and filamentous algae.

Desmids : with cell-wall in two sections which adjoin at the mid-region forming "semi-cells" with protoplast correspondingly divided.

Examples : **Cosmarium Closterium, Xanthidium, Euastrum**

Filamentous green algae : These are branched or unbranched multicellular forms.

Examples : **Spirogyra, Oedogonium, Ulothrix, Zygnema**

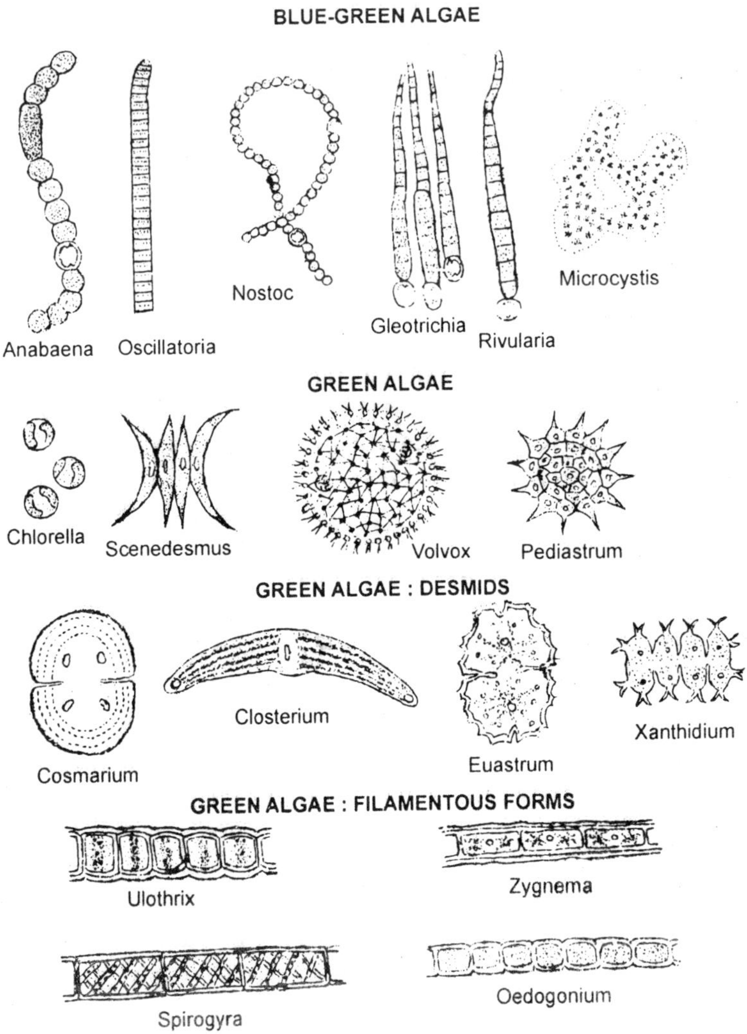

Fig. 11

General Characteristics

1. Unicellular with cells-walls
2. Capable of photosynthesis (Chlorophyll a & b)
3. Chloroplasts present.
4. Flagella present; usually two (one long, one short)

Examples : **Euglena, Phacus, Tranchelomonas, Lepocynclis**

Charophyta (stonewort) : Erect, branched thallus differentiated into a regular succession of nods and internodes.

Bacillariophyta (Diatoms) : It's an important group of phytoplankton.

General Characteristics

1. Golden-brown in colour (carotenoid pigments)
2. Unicellular, sometimes adhere to one another forming chains.
3. Cell-wall of silica (frustules), with minute clear markings composed of two valves.
4. Each diatom cell contains one nucleus.
5. Reproduction usually vegetative; sexual reproduction by auxospore formation.

Two groups of diatoms :

(a) Centric type : Exhibits raidial symmetry

Example : **Cyclotella, Stephanodiscus**

(b) Pinnate type : Exhibits bilateral symmetry

Examples : **Navicula, Amphora Pinnularia, Skeletonema, Fragilaria (Fig. 12)**

CHRYSOPHYTA

A. Chrysophyceae (golden-brown algae)

1. Chromatophore golden-brown colour due to β -carotene and xanthophyll carotenoids; have chlorophyll 'a'.
2. Unicellular or colonial; rarely filamentous
3. Flagella : one or two of equal length

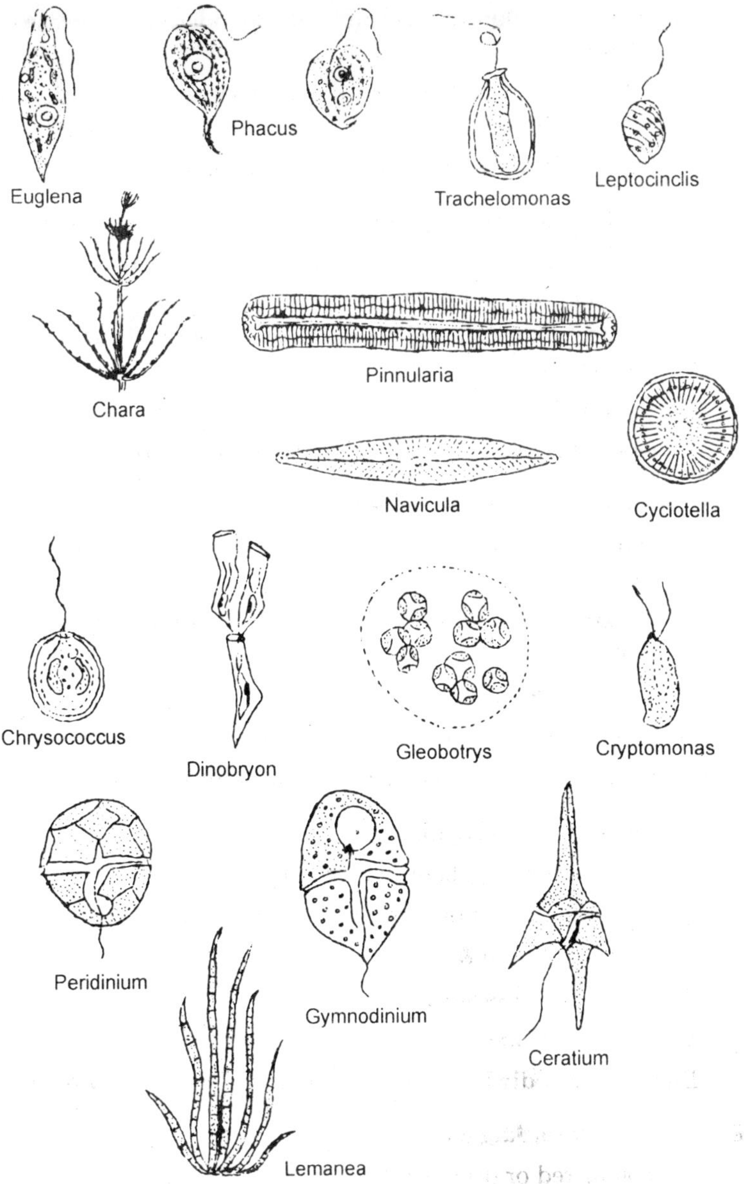
Euglena
Phacus
Trachelomonas
Leptocinclis
Chara
Pinnularia
Navicula
Cyclotella
Chrysococcus
Dinobryon
Gleobotrys
Cryptomonas
Peridinium
Gymnodinium
Ceratium
Lemanea

Fig. 12

4. Many without cell-wall and only bounded by cytoplasmic membrane
5. Vegetative reproduction by longitudinal cell division is most common.

Examples : **Chrysococcus, Dinobryon, Synura.**

B. **Xanthophyceae (Yellow-green algae)**

1. Yellowish-green chromatophores; chlorophyll a and c, β – carotene and xanthophyll
2. Unicellular, colonial or filamentous. Two flagella; one longer than the other
3. Cell-wall is often absent; starch is never formed.

Examples : **Gloeobotrys, Chlorobotrys, Gloechloris**

C. **Cryptophyceae (Cryptomonads) :**

1. Naked, unicellular and motile cells; two flagella of equal size.
2. Chlorophyll a and c and α and Σ –carotene; biliproteins presents.
3. Stored food : starch
4. Reproduction by longitudinal division

Examples : **Cryptomonas, Rhodomonas, Chromonas**

D. **Phyrrophyta (Dinoflagellates)**

1. Unicellular, flagellated and motile
2. Cell-well absent or if present, composed of cellulose
3. Chlorophylla a & c
4. Food stored as starch
5. Mostly marine

Examples : **Peridinium, Gymnodinium, Gonyalaux, Ceratium**

E. **Rhodophyta (red algae)**

1. colour red or dull purplish green
2. mostly marine forms; Order Florideae : freshwater
3. multicellular thalli; cell-wall contains cellulose and protein

4. chlorophyll a & b; blue (phycocyanin) and red (phycoerythrin) pigments
5. Reserve food : Floridean starch

Factors Affecting Growth of Phytoplankton

A. Light and temperature

Growth of algae and photosynthetic rate are directly related to quantitative light intensity. Algae can tolerate wide variations in temperature.

1. *Adaption of light intensity*

a. Some algae response to light intensity by increasing and decreasing the amount of pigment per cell. Cells adapted to high light intensity have a lower chlorophyll 'a' content than those adapted to low light intensities e.g. **Chlorella.**

b. In some other algae, adaptation to changing light intensity occurs only changing the rate of light saturation of photosynthesis. Chlorophyll content is the same in cells at low and high light intensities, e.g. diatoms.

c. Some algae are intermediate between the above two groups of algae.

2. **Adaptation to Low Temperature.**

Some algae adapt to low temperature by increasing enzyme concentrations so that the same rate of photosynthesis is achieved as per given light intensity at high and low temperatures, e.g. **Skeletonema.** In the diatom **Skeletonema,** adapted to temperature, the amount of protein per cell was twice as much as at 7°C as at 20°C and the size of cell is greater at low temperature.

Light of high intensity is distinctly detrimental to many algae. Ultra-violet light is particularly harmful to planktonic algae.

B. Floating Mechanism and Water Turbulence

1. Floating mechanism (menan, to remain within the photic zone).

(a) Adaptations to improve floatation or reduce sinking rates :

Density of most freshwater plankton is generally from 1.01 to 1.03 times of water. This causes them to sink. According to Stoke's Law, "Spherical particles of a mean diameter of not more than 0.5 mm sink."

(b) Various characteristics to improve flotation :

(i) Increasing surface-to-volume ratios by either elongation into cylindrical or discoidal shapes, or by projection of species or other cellular processes.

(ii) Production of mucilage (density of mucilage is less than the cell).

(iii) Gas vacuoles : These are gas-filled structures in protoplasm of living cells.

(iv) Accumulation of fat.

2. Water turbulence (critical in the basic regulation of algal peiodicity and production).

Importance of water movements in the transport of particulate organic matter is well recognized. These are :

(a) Physical movements of algae into or out of the photic zone.

(b) Vertical transport of mineralized matter from lower depths and littoral regions to the open water.

C. **Inorganic Nutrient Factors**

Importance of phosphorous, nitrogen and other inorganic macro- and micro-nutrients for plant growth is well-known. **Liebig's Law of Minimum** : Each organism requires a certain number of food materials and each of these materials must be present in a certain quantity. If one of these food substances is absent, the organism dies; if not absent but present in minimal quantity, the growth will be minimal. This happens even though ample amounts of other required substances are present. According to this law, "Yield of any organism is determined by the quantity of that particular substance that in relation to the needs of the organisms is least abundant in the environment."

D. **Organic Matter and Algal Growth**

Many algae can supplement primary photosynthetic autotrophy by the uptake and utilization of organic substrates. They are capable of synthesizing organic growth factors such as vitamins. They are termed as **Auxotrophic.** Some need light energy for cell metabolism and are known as **Photo-autotrophs** while other termed **Mexotrophs** which also assimilate CO_2 in small amounts along with organic compounds both in light and especially in dark. A few algae termed **Heterotrophs** sustain growth and cell division in the dark; energy and cell carbon are obtained from metabolism of organic substrates.

E. **Biological Factors**

Biological factors affecting growth of phytoplankton are :

(a) those which complete for essential available resources;

(b) micro- or macro-organisms predating on algae.

In short : "Each species of algae possesses a range of tolerance to extreme of these factors and growth proceeds best at a given optimal combination of the interacting factors."

17

PRODUCTIVITY OF FOOD IN LAKES

Standing Crop : It is the weight of organic material that can be sampled or harvested at any one time from given area.

Biomass : It is the weight of all living material in a unit area at a given time.

Yield : It is the crop expressed as a rate (kg/ha/yr.)

Primary production : It is the energy accumulated by plants in photosynthesis.

Primary productivity : It is the rate of energy storage (per unit of time) by plants or it is defined as the rate at which radiant energy is stored by photosynthetic and chemosynthetic activity of producer organisms (mainly green plants and also some bacteria).

Production Process : Basically there are four steps :

A. **Gross Primary Productivity** (also "Total photosynthesis" or "Total assimilation") : It is the total rate of photosynthesis including the organic matter (energy) used up in respiration.

B. **Net primary productivity** (also, "apparent photosynthesis" or "Net assimilation") : It is rate of storage of organic matter in plant tissues in excess of the respiratory utilization by the plants.

C. **Net community productivity** : It is the rate of storage of organic matter, that is, those not used by Heteotrophs.

D. **Secondary Productivity** : It is the rate of energy storage at consumer's level (equivalent to assimilation).

How to Measure Primary Productivity?

Phytoplankton productivity is summarized in the equation :

$$6CO_2 + 6H_2O \xrightarrow[\text{(Pigment receptor)}]{\text{Light}} C_6H_{12}O_6 + 6O_2 \quad \text{(Glucose)}$$

Methods of Measurement

1. Direct measure of the energy flowing through the system (Ideal but difficult).
2. Indirect measure :

   (a) Amount of substance produced.

   (b) Amount of raw material used.

   (c) Amount of by-product released.

A. Light and dark bottle **oxygen** method :

   (Determination of the amount of O_2 produced over a time period).

This method is quite simple and gives a measure of both photosynthesis and respiration at the same time. Three BOD bottles (300 ml each) are taken of which two are clear (light bottles) and one is made 'dark' by painting black and wrapped with aluminum foil (dark bottle). The three bottles are then filled with water samples from the lake/pond containing phytoplankters. Two of the bottles, one clear (LB) and one dark (DB) are next resuspended in the same lake/pond water from non-shading float at the depth from which the initial samples were taken, for a period of 1 to 12 hours (day time). The O_2 -content of the third clear bottle (IB) is immediately determined (Winkler's Method). At the end of the incubation period the O_2 -contents of the other two bottles are also determined.

C_1 = initial concentration of O_2 ppm/hr. .... (IB) 6 ppm

C_2 = Final concentration of O_2 in dark bottle (O_2 ppm/hr) .. (DB) 5 ppm

C_3 = Final concentration of O_2 in light bottle (O_2 ppm/hr.) .. (LB) 10 ppm

So, net photosynthetic activity = LB - IB (C_3 - C_1) 10-6 = 4 ppm

Respiratory activity = IB - DB (C_1 - C_2) (per unit 6-5 = 1 ppm vol. over time of incubation).

Gross photosynthetic activity = LB - DB (C_3 - C_2) or 10-5 = 5 ppm

(Net PS Act + Resp. Act) (C_3 - C_1) + (C_1 - C_2)

Discrepancies of this Method

(a) Respiratory rate not necessarily the same in LB and DB. This method does not measure the community at the bottom.

(b) The smallest amount of photosynthesis : This method can measure under ideal conditions is about 20 mg C m^{-3} day $^{-1}$.

Error ± 20 mg C m^{-3}

To convert ppm O_2 to ppm C fixed per cu. m., the following equation is used :

mg C^{-hr}/cu. m. = ppm O_2 × 12/32 × 1000.

The 14 C technique is very useful in measuring the primary productivity of waters of low productivity. Here a known amount (small) of radio-active 14 CO_2 is added to an air-tight container with phytoplankton in it. the radioactive 14 C is assimilated by photosynthesis similar to non-radioactive C in the containers. The proportion of radio-active to non-radioactive C assimilated is similar to the proportion in the source. By measuring the radioactivity of the phytoplankton after a certain period the rate of carbon fixation can be calculated (by dividing the amount of radio-active 14C by its proportion of the CO_2 in the container at the beginning of the experiment).

Problems

Over long periods the plant's respiration may release some of the assimialted 14C back into the water as CO_2. for shorter period (1-3 hours) the 14C uptake estimates the gross productivity. However, the respiration rate in tropical waters is high and this often resulted in very low values of net production.

The limit of sensitivity of the 14C method is some 5-100 times greater than O_2 method, on the order of 0.1-1 mg C m^{-3} day.

Vertical Migration of Plankton

A few terms

Photic zone : Lighted region of lake. A well-formed epilimnion usually designated as the photic zone.

Aphotic zone : Region of light deficiency. In this zone photosynthesis remains below compensation level or none.

Dysphotic zone : Twilight zone (Region in between photic and aphotic zone)

Euryphotic zone : An organism that has wide range of tolerance in respect to light.

Stenophotic : An organism with a narrow range of tolerance in respect to light.

Seasonal Patterns and Periodicity of Algal Biomass

General trend in lakes (Polar, temperate and tropical)

(a) Successional seasonal periodicity of phytoplanton biomass is reasonably constant from year to year.

(b) Seasonal amplitude of changes in numbers and biomass of phytoplankton is usually very great. Temperate and Polar (FW) lakes : Thousand-fold. tropical waters : Five-folds and above.

In Polar and Temperate FW lakes, however, there is distinct-periodicity in biomass of phytoplankters.

(A) **Winter** : Low light and temperature; growth greatly reduced or negligible; rapid decline in population.

(B) **Spring** : Improved light conditions; increase in phytoplankton numbers and biomass reach maximum; predominant groups; diatoms.

(C) **Summer** : In Eutrophic lakes : brief summer minimum followed by late summer profusion of blue-algae. In Oligotrophic lakes : Low populations.

(D) **Autumn** : Development of second maximum : diatoms.

Periodicity of Species Composition

Seasonal community consists of a composite of :

(a) Holoplanktonic (perennial) species : Throughout the year.

(b) Meroplanktonic (intermittent) species : These enter into some type of diapause or resting stages.

Winter populations : Adapted to cold-water and low light intensity; usually small and motile algae.

Example : **Cryptomonas, Gymnodinium, Dinobryon, Chlamydomonas, Synedra, Fragilaria** and **Trachelomonas.**

Spring populations (maximum) : Mixing of nutrient-laden water; increasing light; dominated by : diatoms.

Spring Decline : Reduction of nutrients in photic zone; decline in diatom population; succeeded by green algae and later by blue-green algae.

Summer populations : Dominance of Blue-green algae.

Aquatic Macrophytes

A. Rooted and attached to substratum

1. Emergent : **Phragmites, Typha, Scirpus, Carex, Cyperus, Juncus Eleocharis, Sagittaria,** grasses and sedges.
2. Floating-leaved macrophytes : **Nymphaea, Nelumbo, Nuphar, Nymphoides, Potamogeton, Victoria**
3. Submerged macrophytes : **Chara** and **Nitella** (algae), isoetes (fern), **Hydrilla, Anacharis, Cabomba, Najas, Vallisneria, Otellia, Myriophylum**

B. Macrophytes not attached to substratum :

1. Freely floating macrophytes : **Eichhornia, Pistia, Azolla** and **Salvinia (ferns), Spirodela, Lemna, Wolffia**
2. Submerged : **Ceratophyllum, Utricularia**
3. Marginal plants spreading over water surface : **Ipomea, Jussiaea, Enhydra, Marsilea (Fern)**

18

ZOOPLANKTERS

Phylum : Protozoa

General characteristics : Microscopic, single called or in colonies of similar cells; without any tissues; Protozoas feed on detritus, bacteria, fungi, yeast and other Protozoans.

(A) **Sub-phylum : Plasmodroma** : Organelles for locomotion are flagella. Pseudopodia or non-nuclei of one kind.

(1) **Class : Sarcodina** : Pseudopodia for locomotion, e.g. **Amoeba, Difflugia, Actinophrys, Arcella**

(B) **Sub-phylum : Ciliophora** : Cilia or sucking tentacles in at least one stage of life history; nuclei of 2 kinds

(1) **Class : Ciliata** : having cilia or locomotion e.g. **Coleps, Stentor, Paramoecium, Vorticella, Lacrimaria, Stylonychia**

Phylum : Rotifera

Characters

1. Microscopic forms (40 µm - 3 mm); body distinguishable to head, trunk and foot.
2. A cliated corona in the head; mouth located in corona (wheel).
3. Trunk covered by cuticle; in some thickened to form lorica (rigid).

4. Foot telescopically retractable; bears 1-4 toes.
5. Pharynx muscular (mastax) having hard jaw (trophi) for sucking and grinding food particles.

Feeding Habits of Rotifers

1. Detritus Feeders : Feeding by means of ciliary movement of corona directing food to mount e.g. **Brachionus, Keratella, Lecane.**
2. Carnivores : They feed by grasping and trapping; trophi used as forceps; presy seized, macerated, edible parts sucked in and rest discarded. e.g., **Asplanchna**

Reproduction in Rotifers :

(A) **Uniparent Reproduction** (Under favourable conditions) :

Diploid females produce diploid eggs which hatch into more diploid females parthenogenetically.

(B) Two parent reproduction (Under unfavourable conditions)

Under stress of crowding, temperature, photoperiod or diet changes males or haploid eggs are produced and sexual reproduction occurs. The fertilized resting eggs produced, have greater genetic viability to survive adverse conditions in the following years.

Phylum : Arthropoda (animals having jointed legs)

1. **Class : Crustacea** : Hard shells; two pairs of antennae present.

A. **Sub-class : Brahiopoda** - legs are flattened and leaf-like; with gills as the chief respiratory organs.

(i) Order : Anostraca - no carapace e.g. Fairy shrimps and brine shrimps.

(ii) Order : Nostoscraca - carapace low. e.g. Tadpole shrimps

(iii) Order : Conchostraca - carapace bivalved e.g. clamp shrimps

(iv) Order : Cladocera - minute, usually bivalved carapace e.g. water fleas.

B. **Sub-class : Ostracoda** - minute bivalved, compressed and enclosing the entire body. e.g. **Cypris, Cypria**

C. **Sub-class : Copepoda** - mostly microscopic; a free trunk somites; no appendages in last 4 somites.

(i) Order : Calanoida - e.g. **Clanus** (exclusively marine), **Diaptomus**

(ii) Order : Harpacticoida - e.g. **Harpacticus**

(iii) Order : Cyclopoida - e.g. **Cyclops.**

CLADOCERA (water fleas)

Characters :

1. Minute mostly 0.2-3 mm in size (except **Leptodora** - 12 to 18 mm long).
2. Usually bivalved, not enclosing head, paired eyes fused, median and sessile.
3. Trunk appendanges 4 to 6 pairs; second antennae enlarged for swimming jerkingly.

Feeding : Five pairs of legs on the ventral part of thorax produce a constant current of water and force food along with the water through the valves. Chief foods are algae and protozoa; also organic detritus as well as bacteria.

Reproduction and Development

(A) **Parthenogenetic reproduction :** (Please see reproduction in rotifers - Uniparent reproduction) : Females produce eggs which develop into parthenogenetic females without fertilization. Number of eggs produced per clutch varies [2 (Chloridae) to 40 (Daphnidae)]; eggs are deposited in a brood chamber or pouch; eggs develop in the pouch and hatch in a small form similar to the parent.

(B) **Sexual reproduction :** (Bi-parent reproduction)

Under unfavourable conditions, sexual reproduction occurs.

COPEPODA

Characters of free-living copepoda

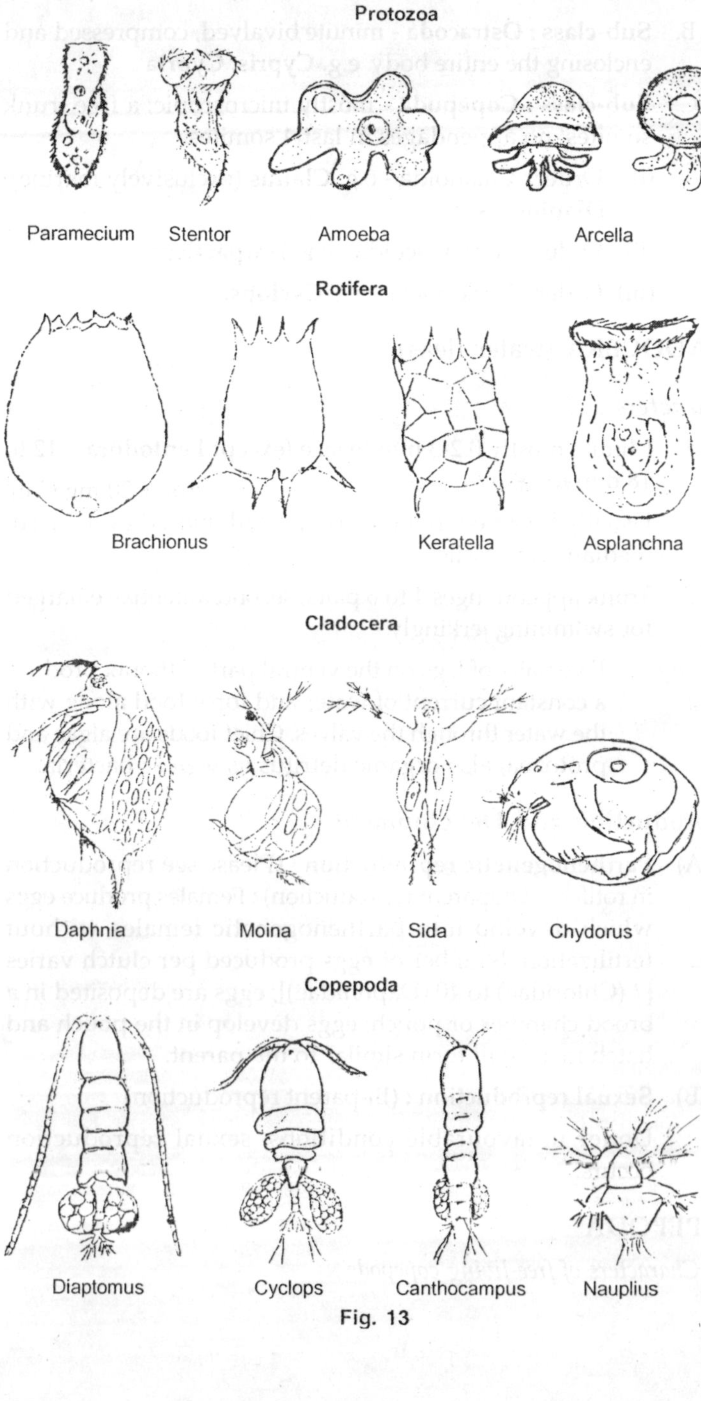
Protozoa
Paramecium
Stentor
Amoeba
Arcella
Rotifera
Brachionus
Keratella
Asplanchna
Cladocera
Daphnia
Moina
Sida
Chydorus
Copepoda
Diaptomus
Cyclops
Canthocampus
Nauplius

Fig. 13

Body consists of **metasomo** (cephalothorax) and **urosome** : head with 5 pairs of appendages (antennae and mouth parts); thorax with 6 pairs of swimming legs; Abdominal segments (in females first segment is modified as genital segment) and terminal caudal rami bearing setae.

Feeding :

(1) Calanoids feed chiefly by filtration of plankton, the antennae being used to produce a current.

(2) Cyclopoids have the mouth parts modified for seizing and biting. Food consists of unicellular plants and animals and also organic debris, e.g. carnivorous forms : **Mesocyclops,** Cyclops. Herbivorous forms : **Eucyclops, Acanthocyclops.**

(3) Harpacticoid mouth parts are adapted for raking, seizing and scraping food from the bottom.

Reproduction and Development

Bisexual (both males and females) : In males the antennae, and in some genera the modified 5th legs are used in clasping the females for the transfer of spermatophores to ventral side of female genital segment. Fertilization may be immediate or even 2 months after copulation.

Fertilized eggs carried by females in egg sacs. Egg no. up to 72 (cyclopid) and 1-30 (calanoid). Higher clutch size in productive lakes (due to food availability). Temperature determines rates of eggs production.

Life History of Copepods

Eggs → Nauplii (full swimming larvae) → molt into **six** Nauplian stages → to first copepodite instar (5 stages during which more appendages and body segments develope) → 6th copepodite stage is the **adult**. Duration of larval stages highly variable

Distinguishing Characters of Three Free-living Copepoda Groups

	Characters	*Calanoida*	*Cyclopoida*	*Harpacticoida*
1.	Anterior part body	Much broader than posterior	Much broader than posterior	Little broader than posterior
2.	Constriction between somite of 4th and 5th legs	Marked constriction bet. 5th & genital leg	Marked constriction bet.	Slight or no constriction
3.	Egg sac	One : medially placed	Two : carried laterally	Usually one : carried medially
4.	First antennae in females	Long : 23-25 segments	Short : 6-17 segments	Very short 5-9 segments
5.	Fifth leg	normal	Vestigial	Vestigial
6.	Habitat	Planktonic; rarely littoral	Littoral; a few sp. planktonic	Exclusively littoral on macrovegetation & sediments.

Phenomenon of Cyclomorphosis

I. **Cyclomorphosis in Rotifers.** (seasonal polymorphism : some part of the organism changes markedly).

(a) **Elongation in relation to body width** : In certain rotifers, the mid-summer population can be 5 times as long as wide; nearly spherical in spring e.g. **Asplanchna.**

(b) **Enlargement with the formation of body wall outgrowths of humps** : This is caused by tocopherol content of plant derived food e.g. **Asplanchna sieboldi.**

(c) **Reduction of size at higher temperature in summer** : Reduction of spine length increases sinking speed in summer e.g. **Keratella**

(d) Production of lateral spine, Posterior spine elongates in **Brachinionus calyciflorus** in presence of the predator. **Asplanchna.**

B. **Cyclomorphosis in Cladocera**

Cyclomorphosis in cladocera occurs more conspicuously than any other group. In **Daphnia** it occurs with :

(a) **Increasing water temperature** (in spring) : A gradual extension of anterior part of head to form a crest or helmet. Also, there is marked decrease in number of parthenogenetic eggs per individual.

(b) **Water turbulence** : Increased antennal beating rate in order to maintain vertical orientation to light. Cyclomorphosis also occurs in **Bosmina, Ceriodaphnia** and **Chydorus;** but much less distinct.

C. **Seasonal cyclomorphosis in Ceratium** (dinoflagellate)

In mid-summer when water gets warm and less viscous, **Ceratium** reduces the rate of sinking out of the photic zone by the reduction of cell-dimension, increased divergence of the horn and by the development of the 4th horn.

VERTICAL DISTRIBUTION OF ZOOPLANKTERS

Since conditions change with increasing depth the vertical distribution of plankers is quite complicated. Generalization is quite difficult.

Tendencies :

a. Greater occurrence of Sarcodine in lower waters.

b. Preference of Dinoglagellates for upper waters.

c. General scattering of Ciliates.

d. Selection of different levels of young and adult stages of certain curstaceans and others.

Vertical distribution is affected by various events of stratification. During periods of thermal and chemical stratifications, certain plankters form distinct concentration of population.

Conditions Influencing Vertical Distribution

(a) light

(b) food

(c) dissolved gases and other substances

(e) wind

(f) gravity

(g) age of individuals.

Vertical Migration of Cladocerans and Copepods

More conspicuous in cladeocera than copepods. Light is the controlling fact for vertical migration; but geotatic reaction also be involved.

Nocturnal migration (upward movement at evening)

Most species migrate upward from deeper waters to more superficial strata as darkness approaches. Single maximum occurs between sunset and sunrise. Vertical convection current also helps vertical migration :

Generally this is due to :

1. Predation pressure
2. Variable food quality of algae; protein synthesis is maximum during night.
3. Growth efficiency is greater at low temperature.

Reverse Migration : (downward movement due to negative phtotaxis) Single surface maximum during daylight period, mroe common under ice-cover (copepods).

Twilight Migration : In this maxima occur surface layers at dawn and dusk.

19
ZOOBENTHOS

Major zones in a lake floor :		Zoobenthos
Littoral Sub-littoral	These are zones of diversity and high productivity	Most types of insects, snails, worms, crustaceans, and fish found. Aquatic plants are abundant in shallow zone.
Profoundal	Below thermocline, uniform physio-chemical factors except decrease in O_2 in summer (Eutrophic lakes	Four main groups : 1. Chironomid midge largae 2. Oligochate worms 3. Chaoborous (phantom midge larvae) 4. Fingernail clams.

SEASONAL CHANGES (EXPECTED) IN BENTHOS OF AN AVERAGE TEMPERATE LAKE OF THE SECOND ORDER

Spring :
(1) Rapid growth of plants.
(2) Periodic insect emergences.
(3) Upward migration of snails (to breed)
(4) Increase in population of various benthic invertebrates.
(5) Beginning stages of summer concentration zone
(6) Migration of **Sialis** larvae to shore.

Summer :
(1) Culmination of plant growth
(2) Fluctuations of population by insect emergences
(3) Shoreward migration by certain mayfly nymphs
(4) Maximum for various chironomid larvae

(5) Continuance for growth and reproduction in various invertebrates.

Autumn : (1) Late autumn decline of shallow-water plants.

(2) General decline of fauna due to insect emergences and scouring effects of autumn storms.

(3) Downward migration of snails, **Sialis larvae** and others.

(4) Persistence of annual summer minimum into early autumn.

(5) Hatching of insect eggs laid in later summer.

Winter : (1) Minimum of plant growth

(2) No insect emergence

(3) Increase in population due to development of autumn-laid insect eggs.

(4) Reduction in reproduction, growth and other physiological processes.

(5) Mid-winter maximum of Tubificidae and Chaoborus larvae.

(6) Gradual reduction of benthos to late winter minimum.

20

FISH

Fishes are cold-blooded aquatic vertebrates which breathe through gills and propel and balance themselves by fins.

They are taxonomically divided into :

(A) **Cyclostomata** : jawless fishes e.g. lampreys and hagfishes

(B) **Chondrichthyes** : cartilaginous fishes e.g. sharks, rays and skates.

(C) **Osteichthyes** : bony fishes carp, perch, etc. majority of living fishes.

Fish as a group has a unique position in lake and stream ecosystems.

Fish have been always neglected by limnologists despite their :

- Ecological importance being at the apex of most aquatic food webs;
- Commercial values, being the main produce harvested from inland waters;
- Fish interact with various levels of the food chains and occupy practically all levels excepting the producers;
- Fish inhabit almost all habitats from highest elevation to the deepest ocean bottom and over 1.5 km in inland Lake Baikal;
- fish have acclimatized to temperatures below freezing point at arctic to over 40°C in some hot springs;
- Fish size vary from the minute (12mm) pigmy goby of hte Philippines to the largest shae shark (22.5m/25mt) and giant freshwater fish of the Amazon River, **Arapaima gigas** (5m/200 kg);

- Fish are the most numerous (over 20,000 species) among vertebrate and have adapted to diverse ecological and environmental conditions.

FOOD AND FEEDING HABITS OF FISHES

Fishes have highly diverse feeding habits involving all different types of feeding; only one broad characteristic prevails, that is, '**The food is taken into the mouth**'. Fish feed on almost all the organisms present in water, such as zoo and phytoplankton, Filamentous algae, macrophyes, benthic organisms; and also feed on fish themselves, amphibians, reptiles, birds and even mammals.

FISH : MAJOR FEEDING TYPES

Predators : Fishes that feed on macroscopic animals.

- Carnivores : that prey on myriad of animal food from zooplankton and insect larvae to large vertebrates e.g. sharks.
- Piscivores : that mainly feed on other fishes e.g. sea-bass, large mouth bass, snake-head (dalag), grouper (Lapulapu)
- Larvivores : that feed on mosquito and other insect larvae e.g. guppy
- Insectivores : feeding mainly on insects e.g. some minnows.

Grazers : Fishes that graze and take by bites often by individual small ones. Food is mainly plankton or bottom organisms, e.g. blue gill

- Planktovores : that feed mainly on zooplanters, e.g. post-larvae (young fry) of almost all fishes; also big head, catla.

Strainers : Fishes that feed by straining microscopic organisms mainly phytoplankton e.g. silver carp.

Suckers : Fishes that such food into the mouth generally from the bottom mud e.g. cotostomids (suckers)

Parasites : Fishes that are parasitic on other fishes; they get attached and such blood and tissue fluid- e.g. sea lemprey.

Other Types of Feeding Habits

Herbivores : fishes that mainly feed on food of plant origin e.g. milkfish

Omnivores : fishes that feed on both plant and animal food, e.g. tilapia, common carp.

Detritivores : that feed mostly on detritus (decaying leaves, twigs, organic debris, fine silt and mud) e.g. Mrigal

Bottom feeders : that feed on bacteria, fungi, protozoans, insect larvae, worms, etc. living on bottom debris, e.g. Mud carp.

Macrophytophagus : that feed on higher aquatic plants, grass, etc., e.g. grass carp.

Malacophagus : that feed on molluscs (snails, etc.), e.g. black carp.

THE BIOMASS PYRAMID IN A FOOD CHAIN

Energy available at each trophic level is one text (approx.) that available at trophic level next below.

Hence biomass at each level is one-tenth (approx.) the biomass at the level next below.

P Photosynthetic algae/phytoplankters are the primary producers.

H Herrings are small herbivorous FISHES

C_1 Mackerels are larger carnivorous FISHES

C_2 Tunas are quite large piscivorous FISHES

C_3 Sharks are very large carnivorous FISHES

C_4 There are huge white sharks which are FISHES and may feed even on man, the top carnivore.

21
Morphometry and Morphology of Lakes and Lakewater

CONTINENTAL LAND MASS

Three hydrological regions:

A. **Exorheic regions:**	From which rivers originate and flow to the Seas, most of the lakes of the world fall under this category
B. **Endorheic Regions:**	Within which rivers arise but never reach the sea, losing themselves in dry water courses or entering closed lake basins. Occur between subtropical deserts and tropical and temperate humid regions.
C. **Arheic Regions:**	Within which no rivers arise, most of the desert areas.

Water balance of lakes:

The sources of water in lakes -

1. Precipitation falling on the lake surface.
2. Water in surface influents.
3. Groundwater seeping through the floor of the lake.
4. Groundwater entering through springs.

Drainage Lakes:	In these lakes, loss of water occurs by flow from an outlet such as a river or stream.
Seepage Lakes:	Loss of water occurs through seepage into the groundwater through the basin walls.
Closed lakes:	These lakes have no outflow effluents and lose water only by evaporation.

Open lakes: Those that have an outflow by an outlet or lose water through seepage.

CLASSIFICATION OF LAKES: BY FOREL (MODIFIED BY WHIPPLE) Table 5

I. Polar Lakes - Surface Temp. ever above 4°C.

Order 1.	Bottom water at 4°C throughout year;
Order 2.	Temperature of bottom water varies but not far from 4°C.
Order 3.	Temperature of bottom very similar to surface water;

II. Temperate lake - Surface temp. varies above and below 4°C.

Order 1.	Bottom water at 4°C throughout year;
Order 2.	Temperature of bottom varies but not far from 4°C.
Order 3.	Temperature of bottom very similar to surface water;

III. Tropical lake - Surface temp. always above 4°C.

Order 1.	Bottom water at 4°C throughout year;
Order 2.	Temperature of bottom varies but not far from 4°C.
Order 3.	Temperature of bottom very similar to surface water.

Heat Budget of Lakes

Temperature changes in lakes are the results of two sets or processes:

Table 5 : Classification of Lakes

Sl.No.	*Characters*	*Eutrophic*	*Oligotropic*	*Dystrophic*
1.	Depth	Relatively shallow	Very deep	Usually shallow
2.	Low water temperature in hypolmnion	Minimal or absent	Present	Variable
3.	Organic matter in bottom and suspension	Abundant	Very low	Abundant
4.	Electrolytes	Variable, often high	Low or variable	Low
5.	Calcium, Phosphorous and Nitrogen	Abundant	Relatively poor	Very scanty
6.	Humic materials	Slight	Very low or absent	Abundant
7.	Dissolved oxygen in deeper water	Minimal or absent	High at all depths	Almost absent
8.	Larger aquatic plants	Abundant	Scanty	Scanty
9.	Plankton : Quantitatively	Abundant	Restricted	Variable (commonly low)
10.	Plankton : Qualitatively	Variable	Rich	Poor
11.	Water blooms	Common	Rare	Variable
12.	Predominant Phytoplankton	Blue-green algae & diatoms	Green algae	Blue-green algae very rich
13.	Profoundal fauna : Quantitative/Qualitative	Poor in hypolimnion	Relatively rich	Poor to absent
14.	Fauna type (Benthic)	Chironomous***	Tanytarsus*	Chironomus (sometimes)
15.	**Chaoborus** (Corethra) : Midge (Benthic)	Present	Usually absent	Present
16.	Deep-dwelling cold water fish species	Usually absent	Common to abundant	Absent
17.	Fish production	Usually good	Poor	Very poor
18.	Succession to....	Swamp or marsh	Eutrophic type	Peat bog (last stage of lake)

*Tanytarsus type : Tanytarsus, midge (insects); its larvae are benthic & intolerant to low O_2 levels.

Chironomus type : **Chironomus (tendepes), also a midge; its larvae (bloodworms) are benthic and resident to low O_2 levels.

	Addition of Heat		Loss of Heat
1.	Radiation from sun and sky	1.	Evaporation
2.	Condensation of water vapour	2.	Convection to atmosphere
3.	Convection from atmosphere	3.	Convection to lake basin
4.	Convection from lake basin	4.	Back radiation from lake surface to atmosphere.
5.	Heat from chemical changes		
6.	Heat from friction in water movements		

Gross or Crude heat budget: The amount of heat necessary to raise its water from 0°C to the maximum temperature found in summer.

Annual heat budget (Maximum heat content): The amount of heat necessary to raise its water from the minimum temperature of winter to maximum summer temperature.

Summer heat income: The amount of heat necessary to raise its water from 4°C at the maximum summer temperature.

Winter heat income: The amount of heat that is necessary to raise the temperature from its minimum heat content to 4°C.

SOLAR RADIATION
↓
PROVIDES
↓
HEAT
HEAT DRIVES
↓
WIND ENERGY
WIND ENERGY PROPAGATES
↓
WAVES
WAVES PROVIDE
↓
MIXING FORCES
IN LAKE WATERS

Similarly, Solar Radiation transmits light directly to aquatic environment thus heating lake water.

And also

Powering Photosynthetic Mechanisms of plants.

Solar Radiation consists of:

1. Ultra-violet region - [wave lengths: 300-380 nm]
2. visible region (light) - [wave lengths: 380-760 nm]
3. Infra-red region - [wave lengths: 760-3,000 nm]

LAKES, THEIR ORIGIN, FORMS AND DISTRIBUTION

Definition

Lakes are inland bodies of water that occupy depressions in the surface of the land. The depressions are called basins.

Or

A lake is sometimes defined as a body of standing water completely isolated from the sea and having an area of open, relatively deep water sufficiently large to produce somewhere on its periphery a barren wave-swept shore. (Welch)

Lake basins are formed by:

I. **Changes in the earth's crust:** (Tectonic agent): Tectonic basins are the result of faulting or warping of the earth's crust. They may be:

(a) Those associated with single faults (Abertl. of Oregon)

(b) Those associated with multiple faults or grabens e.g. lake tahoe (cal. usu), L. Baikal Siberia, USSR), L. Biwa (Japan), and also the chain of rift lakes in Africa, such as, L. Malawi, L. Tanganyika, L. mobutu, (Fig. 14)

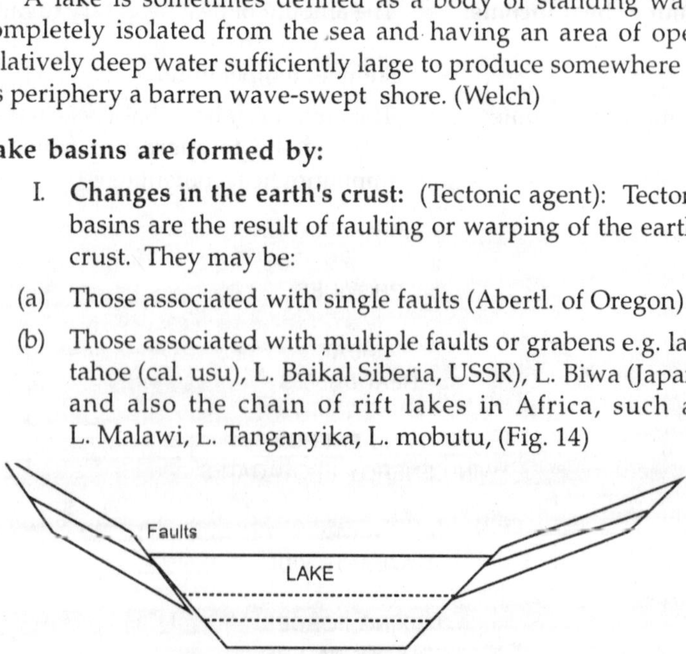

Fig. 14 : Graben Lake (side view)

II. Those arising from the uplift of portions of the sea floor, e.g shallow lake of Florida L. Okeechobee (1,840 km^2), Florida, USA.

Caspian L. and Arel lakes were also created by uplift of the sea bottom or closing of the Tethys sea by plate tectonics. These basins called Relict lakes, retain their original marine structures.

III. Those formed by up warping of earth's crust so as to cause the reversal of the existing drainage system e.g. L. Kioga (from the reversed flow of Kafu river in E. Africa).

IV. Tectonic basins formed by uplifting of the land surrounding the Basin: e.g. lake Victoria (Africa) l. Titicaca in the landes (Peru).

MORPHOLOGY OF LAKES

Form of shore lines in relation to lake origin

A. **Circular:** Crater lakes, calderas and Maars (crater lake), Doline or solution lakes (deep L. Florida) development of shoreline approaches unity in such lakes. 1.04-1.15 in Maars.

b. **Sub-circular:** Cirqe lakes, kettle lakes (cavities left by melting of ice blocks in outward discharge into a valley).

c. **Elliptical:** Oriented lakes of the arctic.

d. **Subrectangular elongate:** Most of the Grabens (L.Baikal, Lake Tanganyika)

e. **Dendritic:** When a high dam is formed flooding the tributary valleys (lake Kioga)

f. **Lunate:** Oxbow lakes and also a few special type of calderas.

g. **Triangular:** Lakes in drowned valleys behind bars (shoreline lakes of landes area, W France)

h. **Irregular:** Regions of glacial scouring of shattered rocks or where the lake is formed by fusion of several basins (loch Trealaval of Scotland)

Insulosity:

Insulosity of a lake is the percentage of area within the shoreline that is occupied by islands.

Lake Huron has the largest islands inside a freshwater area

Elevation of lakes:

Lake Erie:	Above sea level
Lake Superior:	210 m below sea level
Lake Huron:	52 m below sea level
Lake Michigan:	98 m below sea level
Lake Ontario:	163 m below sea level

GEOMORPHOLOGY OF LAKE BASIN

(Origin of lake basins)

How does this study help study of lakes?

- It reflects physical, chemical and biological events in the lake basins.
- It plays major role in lakes metabolism and its control
- It controls nature of lakes drainage
- It controls inputs of nutrients in the lake
- It regulates volume of influx in relation to flushing renewal time.

Shape of a lake basin is reflected in its productivity

- Steep-sided U- or V-shaped basin [formed by tectonic forces] are usually deep and unproductive.
- Shallow basins with greater percentage contact of water with soil show higher productivity
- Basins with longer, irregular shorelines are usually more productive because of greater, contacts of water with soil.

Table 6 : Some Important Lakes of the World

Name	*Continent (Country)*	*Area (km^2)*	*Depth*		*Type of basin*	*Remarks*
			Zm m	*z m*		
Superior	N. America (USA-Canada)	83,300	307	145	Glacial	Largest Freshwater Lake
Michigan	USA	57,580	265	99	Glacial	The Laurentian great lake (area;
Huron	USA - Canada	59,510	223	76	Glacial	242,000 km^2) constitute the
Erie	USA - Canada	25,820	64	21	Glacial	greatest continuous mass of
Ontario	USA - Canada	18,760	225	91	Glacial	freshwater of earth
Great slave	Canada	30,000	614	–	Glacial	
Great bear	Canada	29,500	> 137	–	Glacial	
Tanganyika	Africa	34,300	1,470	572	Tectonic (Graben)	Second deepest
Malawi (Nyassa)	Africa	30,800	706	203	Tectonic (Graben)	in the world
Victoria	Africa	68,800	79	40	Tectonic (Upwarfing)	Wet season : 40,000 km^2
Chad	Africa	16,500±	12±	1.5±	Tectonic	Dry season : 6,000 km^2
Baikal	Asia (USSR)	31,550	1,741	730	Tectonic (Graben)	Deepest lake in the World 20%
Balkhas	USSR	17,575	26.5	6.13	Tectonic	of all freshwaters
Caspian Sea	Asia	436,400	946	182	-do- uplift of sea	Largest lake (Saline)
Aral Sea	Asia (USSR)	62,000	68	15.6	bottom -do- uplift of sea bottom	Saline

Table 7 : Important Lakes of the Philippines

Name	*Location*	*Area*	*Zm m*	*Origin*	*Remarks*
Laguna	Luzon	900 km^2	2.5 m	Tectonic	
San Pablo lakes	(Lagun) LAGUNA	2.9 km^2 (Total)	7.2-156 m	Volcanic (caldera)	consist of seven lakes*
Tadlak	Laguna	0.27 km^2	65 m	— do —	
Taal	Luzon (Batangas)	263.65 km^2	198 m	Volcanic	
Bhui	Luzon (Camaranes del sur)	18 km^2	10 m	By land slides	
Paoay	Luzon (Ilocos norte)	4.4 km^2	4.1 m	Dune lake	
Mainit	Mindanoao (Surigao del norte)	141 km^2	223 m	Tectonic	
Maujan	Mindoro (oriental)	79 km^2	19 m	Volcanic	
Lanao	Mindanao	357 km^2	112 m	Volcanic	(Lava flow)

* 1. L. Bunot, 2. L. Calibato, 3. Mohicup, 4. L. Palacpaquen, 5. L. Pandin, 6. Sampaloc L. 7. L. Yambo

LACUSTRINE ZONATION

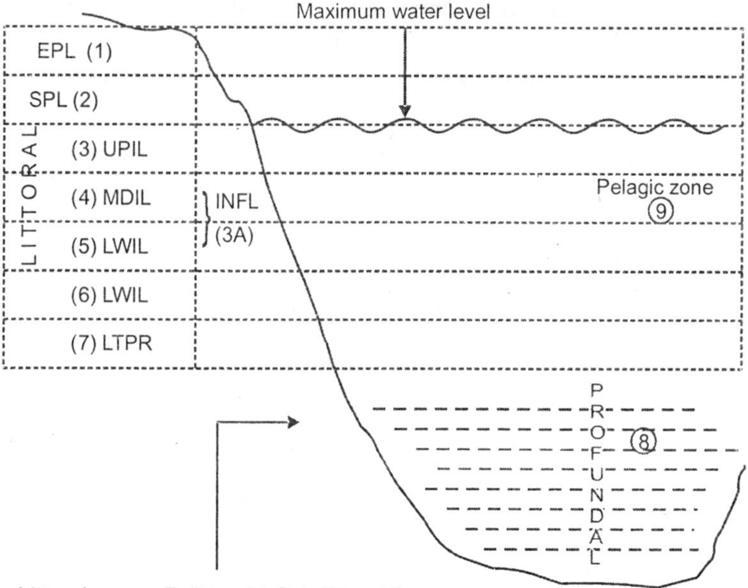

Littoral zone = Eulittoral + Sub-littoral Zones

Fig. 15

Zonation:

1. **Epilittoral zone:** Lies entirely above water level uninfluenced by the spray of the waves.

2. **Supralittoral zone:** Also lies entirely above the water level but is subject to spraying by waves.

3. **Eulittoral zone:** Shore zone between the highest and lowest by the disturbances of …….. waves.

3.a. **Infra or sublittoral zone:** The shoreline region lying between the lowest water level to the lower boundary of macrophytic growth.

This zone (3a) is divided into three sub-zones. They are:

(i). **Upper infra-littoral Zone:** Zone of emergent rooted vegetation.

(ii). **Middle infra-littoral Zone:** Zone of floating - leaved rooted vegetation.

(iii). **Lower infra-littoral Zone:** Zone of submersed footed or adnate macrophytic

4. **Littori profundal Zone:** Zone of scattered photosynthetic forms and includes a gradient of benthic algae, mainly blue-green algae.

5. **Profundal Zone:** Zone comprising of exposed fine sediment free of vegetation

6. **Pelagic Zone:** Free open water of lakes.

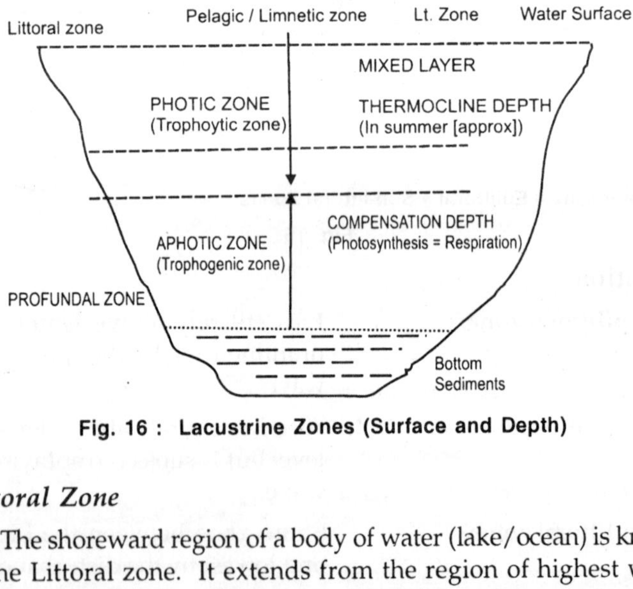

Fig. 16 : Lacustrine Zones (Surface and Depth)

Littoral Zone

The shoreward region of a body of water (lake/ocean) is known as the Littoral zone. It extends from the region of highest water level to a depth where light is barely sufficient for rooted aquatic plants to grow.

Pelagic / Limnetic Zone

(Region of the free water in lakes) In deep lakes, the area beyond the influence of the shore or bottom is termed pelagic or Limnetic Zone.

Trophogenic or Photic Zone

The zone extends from the lake surface to where light is about 1% of that at the surface photosynthetic production occurs at this zone.

Tropholytic or Aphotic Zone

This zone extends below the littoral or photic zone to the bottom of the lake, here light level is too low for photosynthesis. This is a region of O_2 consumption and heterotrophic decomposition of organic matter.

Profundal zone

The deep zone of lakes below the light-controlled limit of plant growth.

Compensation point/Compensation depth

The light intensity at which O_2 production by photosynthesis and O_2 consumption by respiration of the plants concerned are equal, is termed as compensation point and the depth at which it occurs is called the compensation depth.

Some important characters of eutrophic lakes

- Relatively shallow; deep coldwater minimal or absent.
- Organic materials on bottom and is suspension abundant.
- Electrolytes variable, often high; Ca, P and N Abundant; humic materials slight.
- Dissolved oxygen in deeper stratified lakes minimal or absent in hypolimnion.
- Larger aquatic plants abundant.
- Plankton quantitatively abundant, quality variable water blooms common; blue green algae and diatoms predominant.
- Profundal fauna in deeper lakes poor in species and quantity in hypolimnion, chironomus type, corsthra present.
- Deep dwelling coldwater fishes usually absent; suitable for Perch Pine, Bass and other warm water fishes.

EUTROPHIC LAKES
(Rich in nutrients)

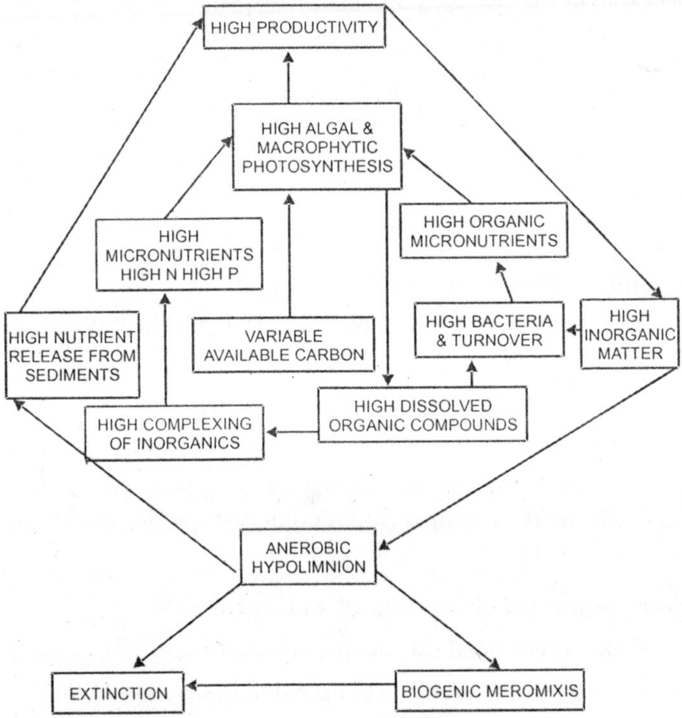

Fig. 17 : Major Inorganic and organic interactions influencing the metabolism of phytoplankton of eutrophic lakes

ORIGIN OF LAKES

2. Volcanic origin: (lakes created by volcanoes)

a. **Crater lakes:** These are formed when a crater of an extinct volcano remains intact and later gets filled with water. They are of 3 types:

(i) Those existing in the unaltered cones of a previous volcano e.g. L. crater butte (California, USA).

(ii) Those formed in the depression resulting from abortive volcanic explosion

These lakes formed at considerable depths below the earth's surface resulting in low rims but deep basin area known as Maars e.g. pulvermaar in Germany (0.35 km^2, 74 m deep) [dia<2 km and even more than 100 m deep].

(iii) Those resulting from the collapse of the rim of a volcano. Collapsed cones which form the largest basins are known as **Calderas** e.g. crater lake (10 km: 600 m deep), oregon, **USA.**

b. Lakes created by damming the outlet of a valley by lava flow:, e.g. snag lake in California, USA.

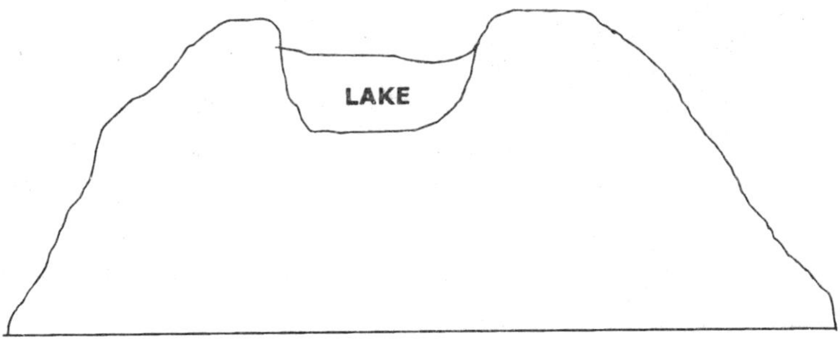

Fig. 18 : Crater lake in a Caldera (minimum dia - 5 km)

3. **Glacial origin** (Lakes formed by glacial erosion)

a. Glacial rock Basins formed by gouging (Digging) Action of a Moving Glacier

(i) Large valley rock basins formed by continental ICE, e.g. great slave lake (Canada); Laurentian lakes of N. America.

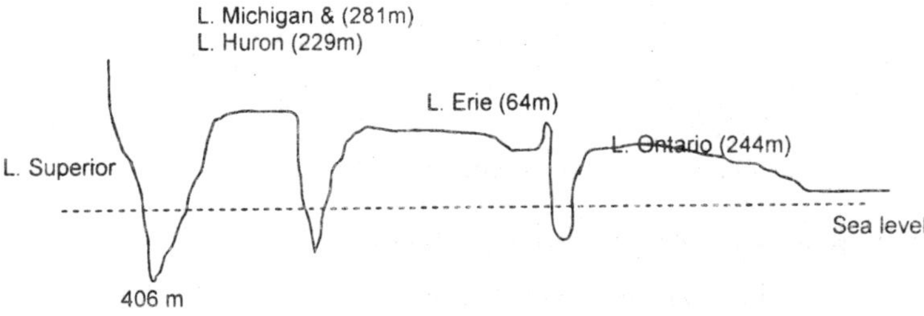

Fig. 19 : Profile of the great lakes (Laurentian)

Origin of lakes

(ii) Lakes which are formed at about the snow line in glaciated valley. These lakes are semicircular and are known as cirque lakes, e.g. Castle lake, California.

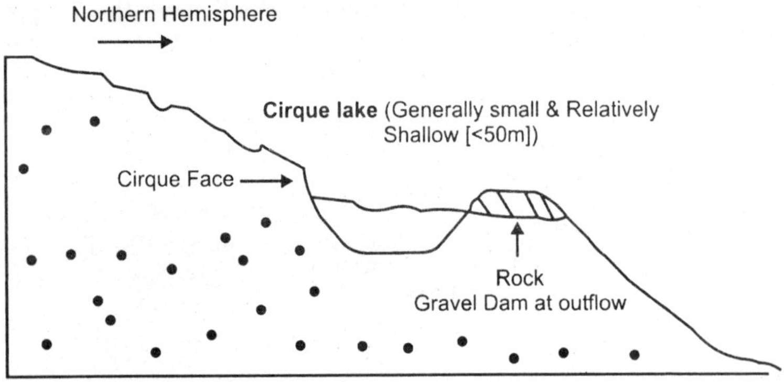

Fig. 20

B. **Drift basins:** Melting of blocks of ICE trapped in glacial moraine form **Kettle lakes**. e.g. Walden pond, Massachusetts (Moraine: Debris carried down & deposited by glaciers) (Fig. 21)

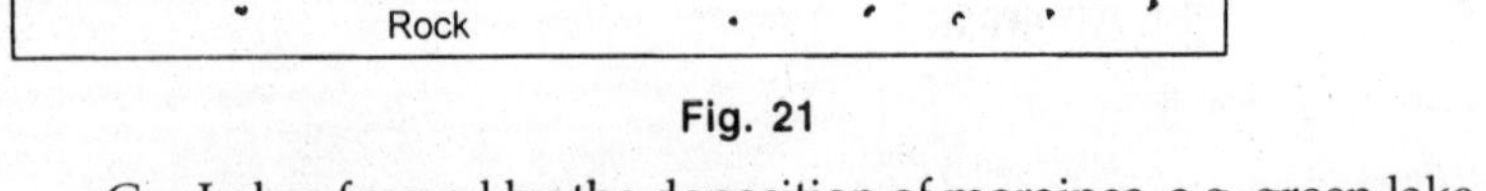

Fig. 21

C. Lakes formed by the deposition of moraines, e.g. green lake, Wisconsin.

D. Lakes formed by Ice or by moraine in contact with existing ICE, e.g. Garshan lake of Central Asia.

4. Lakes with miscellaneous origin

a. Those formed by landslides:

Sudden movement of large quantities of unconsolidated materials in the form of landslides into the floors of a river valley (occasionally initiated by the activity of earthquake), e.g. Lake Buhi, Philippines, L. Sarez, Pamir Mountain.

b. **Solution lakes:** Lakes formed by dissolution of soluble rock/limestone by water, e.g. deep lake Florida.

c. **Fluviatile - Dam lakes:** Those are created by the deposition of sediment by a tributary (running water) across the main river. e.g. Tulare Lake, California>

d. **Lakes formed by flood plains of rivers:** Oxbow lakes are formed when loop of a meander ('S'-shaped winding river) is cut off by silt deposition, e.g. found in the flood plains of most rivers of the world.

e. **Artificial lakes (Man-Made):** A dam built across a river valley will block the flow of water and form a lake e.g. hover dam (R. Colorado), Aswan dam (R. Nile in Egypt) and thousands of other dams.

f. **Lake basins formed by wind action:** Wind action effects redistribution of sand or operates by deflation or erosion of broken rocks to form Dune lakes. e.g. moses L., Washington, L. Paoay, Philippines.

g. **Lakes formed by shoreline activity:** Those formed by building of a sand bar across some irregularity of the cost line, e.g. lakes of landes area on the west coast of France.

h. **Lakes of organic origin:**

    i. Damming due to organic accumulation (deposition of decomposed plants and associated detritus)., e.g. Silver L., Halifax.

    ii. **Beaver dams:** a large numbers of very small lakes in North America have been formed partly or wholly by the activities of beaver.

Manor inorganic and organic interactions influencing the metabolism of phytoplankton of oligotrophic Lakes.

Some important characters or oligtrophic lake:

- Very deep thermocilne high; vol. of hypolimnion large; water of hypolimnion cold.
- Organic materials on bottom and in suspension very low

OLIGOTROPHIC LAKES
(Poor in nutrients)

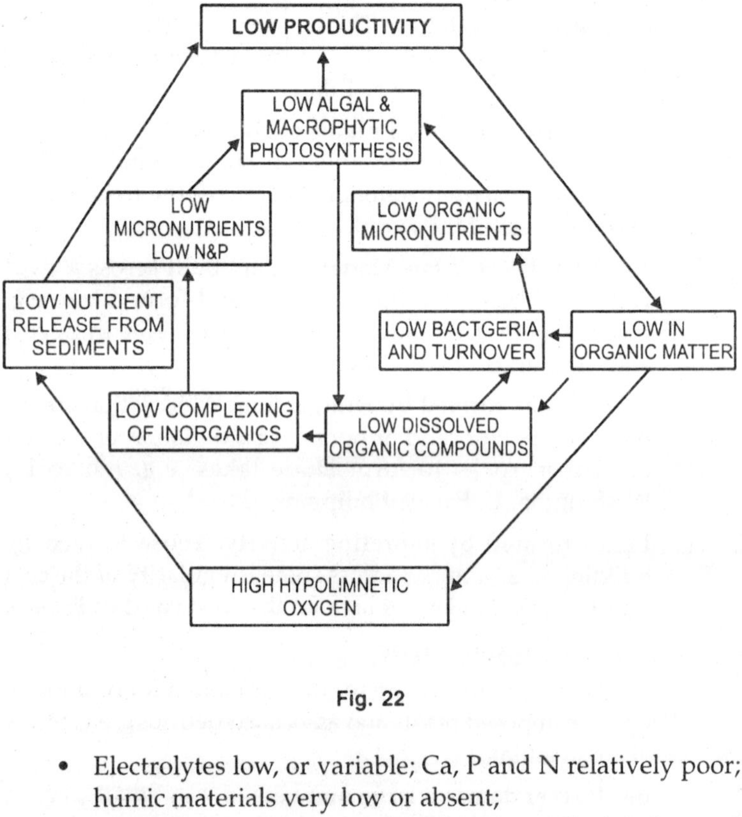

Fig. 22

- Electrolytes low, or variable; Ca, P and N relatively poor; humic materials very low or absent;
- Dissolved oxygen contest high at all depths and throughout the year.
- Larger aquatic plants scanty.
- Plankton quantitavely derestricted; species many; water blooms rare, green algae dominant.
- Profundal fauna relatively rich in species and quantity; tanytarsus type; corethra usually absent.
- Deep-dwelling, coldwater fishes (salmon, trout) common to abundant.

Table 8 : Classification of Lakes

Sl.No.	Characteristics	Eutrophic	Oligotrophic	Dystrophic
1	Depth	Relatively Shallow	Very deep	Usually shallow
2	Low water temperature in hypolimnion	Minimal or absent	Present	Variable
3	Organic matter in bottom and suspension	Abundant	Very low	Abundant
4	Electrolytes	Variable, often high	Low or variable	Low
5	Calcium, phosphorus and nitrogen	Abundant	Relatively poor	Very scanty
6	Humic materials	Slight	Very low or Absent	Abundant
7	Dissolved oxygen in deeper water	Minimal or absent	High at all depths	Almost absent
8	Larger aquatic plants	Abundant	Scanty	Scanty
9	Plankton: Quantitatively	Abundant	Restricted	Variable (commonly low)
10	Plankton: Qualitatively	Variable	Rich	Poor
11	Water blooms	Common	Rare	Variable
12	Predominant phytoplankton	Blue-green Algae and diatoms	Green Algae	Blue-green algae very rich
13	Profundal fauna: Quantitative / Qualitative	Poor in hypolimnion	Relatively Rich	Poor to absent
14	Fauna type (benthic)	Chironomus *	Tanytarsus **	Chironomus (sometimes)
15	Chaoborus (Corethra): Midge (benthic)	Present	Usually absent	Present
16	Deerdwelling coldwater fish species	Usually absent	Common to abundant	Absent
17	Fish production	Usually good	Poor	Very poor
18	Succession into..........	Swamp or marsh	Eutrophic type	Peat bog

** Tanytarsus type: Tanytarsus, a midge (insect); its larvae are benthic and intolerant to low O_2 levels.

* Chironomous type: Chironomus (tendepes) also a midge; its larvae (bloodworms) are benthic and resistant to low O_2 levels.

(Modified from WELCH)

MORPHOMETRIC PARAMETERS

1. *Shore line(L)*

Length of shore line is determined directly or from maps with a map measurer (Chartometer or Rotometer)

2. *Shore line development (DL)*

The ratio of the length of the shore line (L) to the length of the circumference of a circle of area equal to that of the lake

$$D_L = L/2\sqrt{\pi}\,Ao$$

Hypsographic Curve (Depth-area curve):

It is a graphic representation of the relationship between the surface area of a lake and its depth.

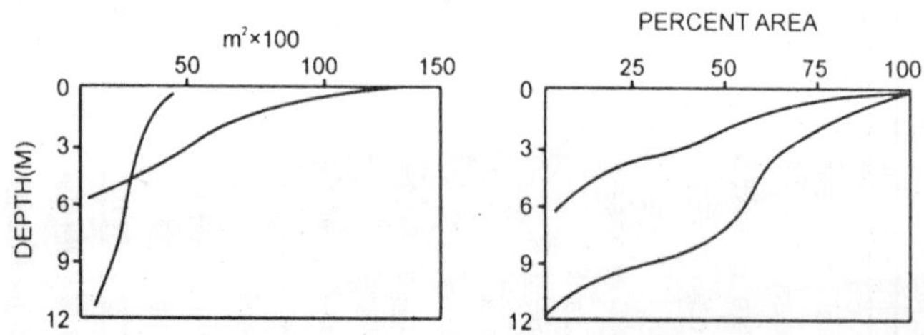

Fig. 23 : Hypsographic depth-area curves of oligotrophic Lawrence Lake and eutrophic wintergreen lake (blue line) in Michigan, USA.

Volume Curve

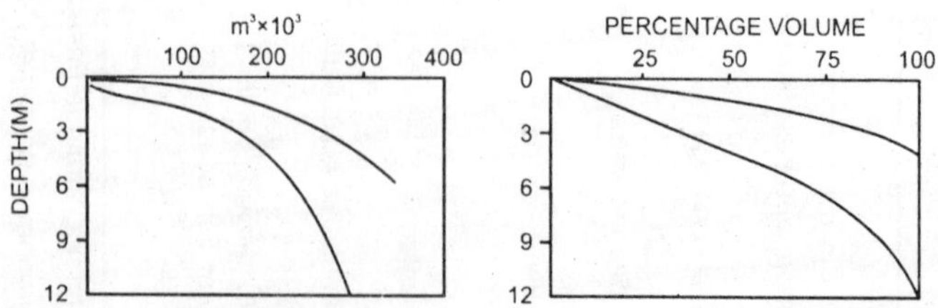

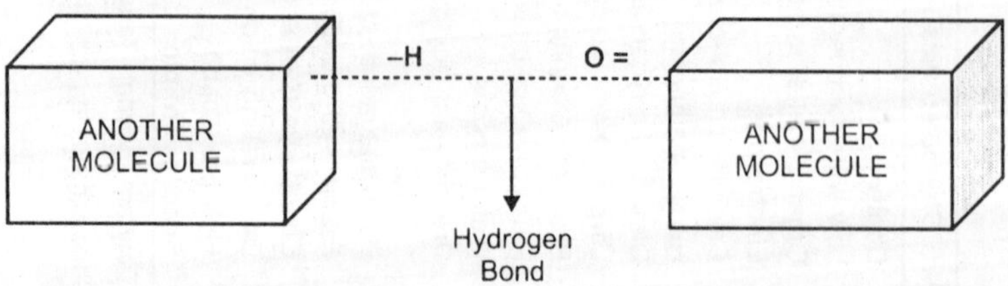

Fig. 24: Depth-volume: Curves of oligotrophic Lawrence Lake (Red line) and eutrophic wintergreen lake (Blue line) in Michigan, USA

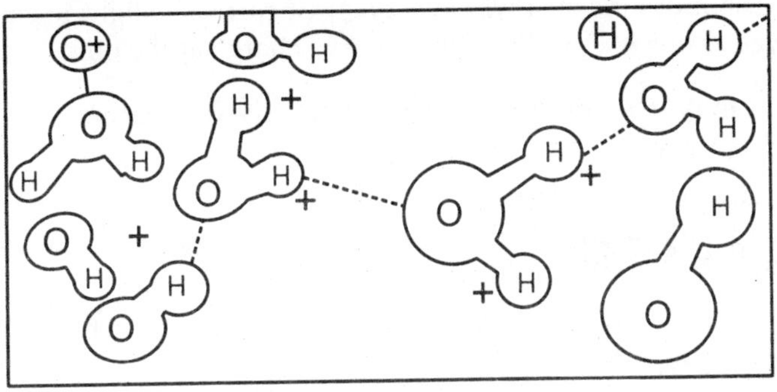

Fig. 25 : Hydrogen bonds between water molecules in liquid water.

In a hydrogen bond, and electronegative atom weakly attracts a hydrogen atom that is covalently bonded at a different atom.

Covalent bond: Where two atoms sphere electrons.

EG H- H [single covalent bond]: 0 = 0 double CVB

Nonpolar CVBs: Both atoms exert same pull on shared electrons e.g. $H:H=H_2$

Polar CVBs: One atom exerts more of a pull on shared electrons, e.g. Electrons shard in a water molecule (H-O-H) are less attracted to the H_2 than to O_2 which has more protons in its nucleus.

Many of the remarkable properties of water arise from this polarity.

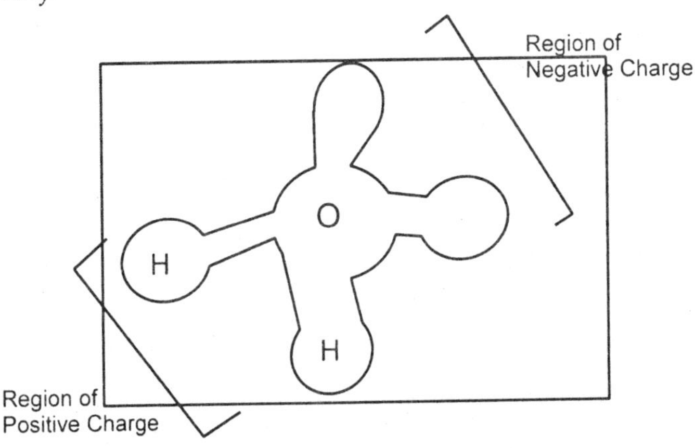

Fig. 26 : Polarity of a water molecule

The molecule as a whole is polar (it carries a slight negative charge at one end and a slight positive charge at the other).

What makes water behave so differently from other liquids or chemicals?

Its molecular structure. This allows weak hydrogen bonding between hydrogen and oxygen atoms in adjacent molecules, producing a matrix known as a liquid crystal.

The maintenance of this liquid crystal gives water most of its unusual properties.

If water behaved at natural environmental temperatures as do H_2S, NH_3 or CH_4, it would be present only as vapour. In fact, water and mercury are the only inorganic liquids that exist at earth's surface under NTP (Normal Temp. and pressure).

The complex bonding holds the water together as a liquid to a much higher temperature than compounds like H_2S, NH_3 or HF.

Water as a universal solvent, transports through biological systems the gases, minerals and dissolved organic components that drive life's machinery.

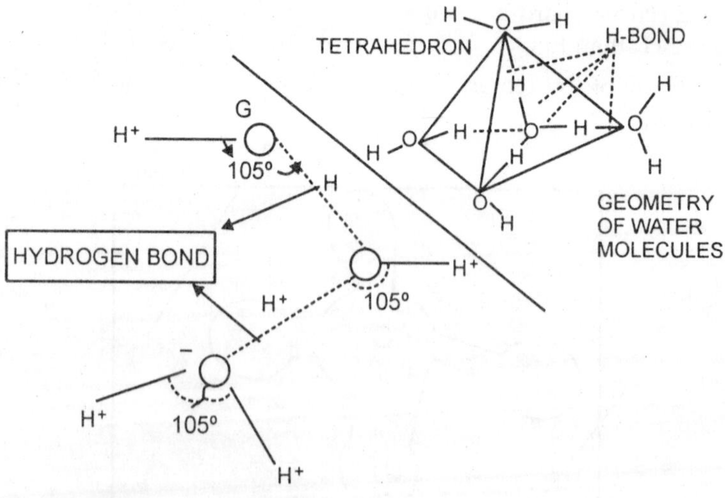

Fig. 27. Hydrogen bonding (dotted lined) in water. The hydrogen bond forms between the hydrogen of one molecule and the oxygen of another. The bond is weak but is more flexible than other bonds.

WATER – VARIOUS FACETS OF QUALITIES

Water's unusual properties such as changes in its density boiling and melting points, viscosity. Surface tension specific heat, high dielectric constant and others have made water a Life-saving compound.

PHYSICAL PROPERTIES OF WATER

Density Varies with : (a) Temperature
(b) Pressure
(c) Dissolved substance
(d) Suspended particles.

Buoyancy: Is a direct outcome of density and varies with the same factors.

Pressure: Water is a heavy substance. 1c.c. weight 1g AT 4°C it increases with depth. Pressure increases one atmosphere with 10m depth. Objects may collapse under heavy pressure of water called implosion.

Compressibility: Water is virtually incompressible.

Viscosity / Mobility: Water is an exceedingly mobile liquid. But it has internal friction. Viscosity is a measure of a liquid's resistance to flow-viscosity of water is 100 times greater than air. It falls with the increase of water temperature.

Table 9 : Changes in the viscosity of pure water due to changes in temperature.

	Temperature °C	*Percentage of viscosity*	
Water has low viscosity as compared to other liquids	0	100	Dissolved salts increases viscosity of water
	5	84.9	
	10	73.0	
	15	63.7	
	20	56.1	
	25	49.8	
	30	44.6	

Surface tension: Water has high surface tension. It occurs due to unbalanced attraction between molecules, since the surface molecules are attracted on one side only. Upward attraction is lacking since there are no water molecules above them. This property is important in the maintenance of protoplasmic form and movement. It acts at the air-water interface and forms a special habitat for organisms adapted to living in surface film (Neuston) organic surfactants (foaming or wetting agents) produced by dissolved salts. Water striders (insects) take advantage of surface tension and skate rapidly on the water surface.

CLASSIFICATION OF ORGANISMS IN WATER

1. Seston: Entire heterogenous mixture of living and non-living bodies which float or swim in water [Sestos: Strainable]

A. Bioseston: Living components of seston

1. ***Plankton:*** Organisms of relatively small size, mostly microscopic having low power of locomotion, dript in water subject to the action of waves, currents and other forms of waters motion [planktons: wandering]

2. ***Neuston:*** Organisms related to surface film of water [neust: swim]

- **a. Infraneuston:** Hanging from or floating against lower side e.g. cladocerans, pond snail.
- **b. Supraneuston:** hanging upon surface film e.g. water striders, water measures.

3. ***Pleuston:*** Higher plants which float either upon the surface or within the water [pleust: float/sail]

4. ***Nekton:*** Organisms of larger size which swim freely regardless of movements of surrounding water [Nektos : swim]

B. Abioseston: Non-living components of seston

1. *Autochthonos detritus:* detritus produced within a given locality [auto: self] [Chtohon: land]

2. *Allochthonous detritus:* detritus originating outside a given locality. [Allos: other]

II. Aquatic biota of long standing: [They are grouped as -]

A. *Littoral:* Organisms inhabiting region adjacent to the shore [Littor: Seashore]

B. *Limnetic:* Organisms inhabiting the open waters

1. True limnetic: organisms thriving in open waters only.
2. Facultative limnetic: organisms which spend part of their life cycle in open waters.

C. *Profundal:* Organisms living in deep region belong the light-controlled limit of plant growth [profound: deep]

D. *Benthos:* Organisms which live in or on the bottom of the basin [benthos: depth].

* Autochthonous detritus: detritus produced by plankton, neuston, pleuston, nekton and benthic organisms.

WATER

Water economy in Lakes

A. Income:

- Precipitation
- Surface influents
- Groundwater sources

B. Losses:

- Outflow from surface
- Seepage to groundwater
- Evaporation
- Transpiration

The hydrological cycle: The cycle consists of three principal phases:

- Evaporation

- Precipitation
- Groundwater runoff

Each phase involves transport, temporary storage, and a change in the state of the water

Global water balance.

HYDROLOGICAL CYCLE

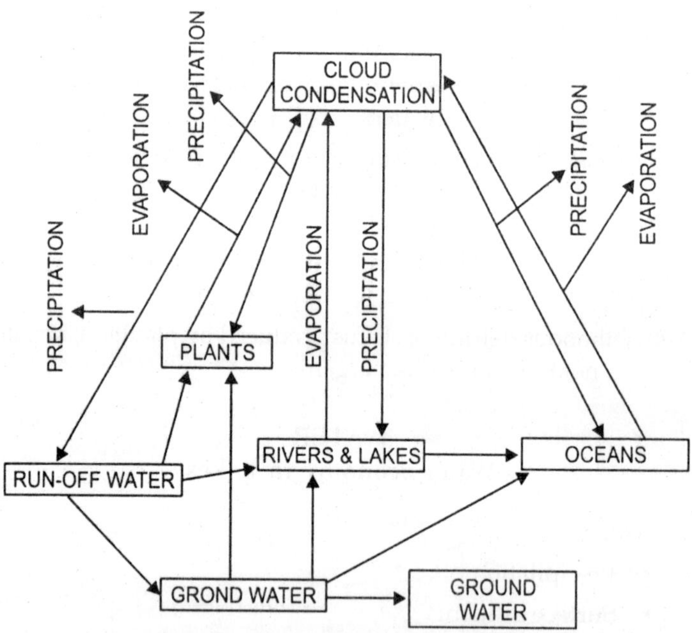

Fig. 28 : Diagram showing phases of hydrological cycle

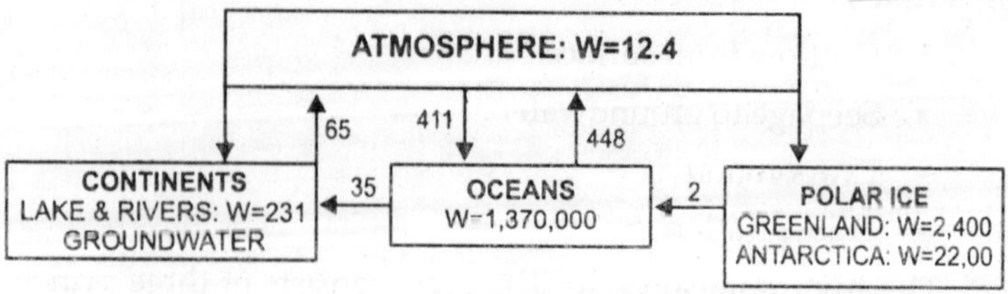

Fig. 29 : Global water balance. W=Water content in 10^3 km^3 values in arrows = transport in 10^3 km^3 yr^{-1} estimate of groundwater is to a depth of 5 km of earth's crust. (Wetzel)

Table 10 : Water in Biosphere

	Volume (1000 km³)	*Percent of total*	*Renewal time*
Oceans	1,370,000	97.61	37,000 yrs
Polar ice, glaciers	29,000	2.08	16,000 yrs
groundwater (deep a.e.)	60,000 4,000	0.29	300 yrs
Freshwater lakes	125	0.009	1-100 yrs
Saline lakes	104	0.008	10-000 yrs
Soil and subsoil moisture	67	0.005	280 days
Rivers	1.2	0.00009	12-20 days
Atmospheric water vapor	14	0.0009	9 days

Water: grand total = 1463311000 km³ or 146 x 1407 km³

[From: J.R. Vallentyne modified]

COLOR OF LAKES

With moderate dissolved or suspended matters the lake appear **Green. Glacial** valley lakes (nearest to glaciers) look white due to suspended glacial flour.

Lake with large quantities of dissolved materials (especially organic) appear yellow or brown.

Phytoplankton blooms produce Green colour. Spectacular lake colors [Red or blue - green] are produced by bloom of near surface blue-green algae.

TURBIDITY

Turbidity is a condition of water resulting from the presence of suspended matters (S.M.). The S.M may be of biogenic (living) or a biogenic origin (non-living).

Settling suspended matters

Settling S.M. are those substances which in still water will sooner or later settle down to the bottom.

Rate of settling depends on the specific gravity, size and by the ratio of surface to volume of the particles and by the viscosity of water.

$$\frac{\text{Sinking velocity of Planktonic organisms}}{} = \frac{\text{Excess of weight}}{\text{viscosity of water} \times \text{form resistance}}$$

Excess of weight = difference in sp. gravity of the body and that of water.

Form resistance = surface area and share of the body.

Non-setting suspended matters

Non Settling S.M. are either those very finely divided solids or those with sp. gr. Less than water and are in permanent suspension.

(a) Planktonic organisms and coarsely divided non living substance whose sp. gr. is such that they are constantly suspended.

(b) Very finely divided non-living materials and organisms of exceedingly small size, such as nanno-plankton.

THERMAL STRATIFICATION

Epilimnion: The turbulent superficial layer of a lake lying above the metalimnion which does not have a permanent thermal stratification.

Hypolimnion: The deep layer of a lake lying **Below** the metalimnion and removed from surface influence.

Thermocline : Metalimnion:

The plane in which the temperature falls most rapidly is called the thermocline and the layers on either side of the thermocline, loosely embracing the whole region of rapid fall in temperature with depth, are called collectively the metalimnion.

The lake of water mixing between the warm epilimnion and the cooler hypolimnion is a major factor determining lake productivity.

[Limne = Lake] [Clino = Slope, inclined]

HEAT: THERMAL PROPERTIES OF WATER

These were discussed in text.

CIRCULATION IN A DIMICTIC HOLOMICTIC LAKE

These were discussed in text.

CLASSIFICATION OF LAKES BASED ON THEIR THERMAL CONDITIONS

These were discussed in text.

DENSITY OF WATER IS MAXIMUM AT 4°C

Water varies	Inversely with temperature
	Directly with pressure
	Directly with dissolved salts and substances
	Directly with suspended particles
	Directly with buoyancy.
Viscosity of water	
Varies	Indirectly with temperature
	Directly dissolved salts / substances
Surface tension of water	
Varies	Indirectly with dissolved salts and substances
Boiling point of water	
	Increases with increase of dissolved salts/ substances
	Increases with increase of pressure
Melting point of water:	
	Decreases with increase of pressure
Freezing point of water	
	Decreases with increase of dissolved salts/ substances increases with increase of pressure

The bicarbonate and carbonate ions dissociate after equilibrium is established:

$HCO_3^- + H_2O = H_2CO_3 + OH^-$ [D]

$CO_3 + H_2O = HCO_3^- + OH^-$ [E]

$H_2CO_3 = H_2O + CO_2$ [F]

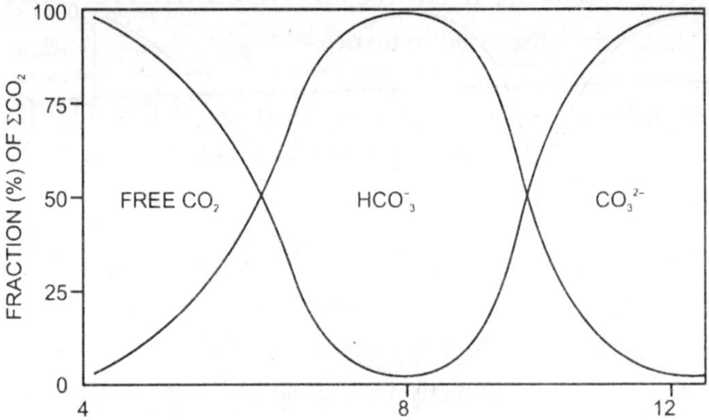

Fig. 30: Relation between pH and the relative proportion of inorganic carbon species of CO_2 [CO_2, HCO_3, CO_3^{2-}] N_2 Solution

What is pH?

Pure water dissociates weakly to H^+ and OH^- ions. The dissociation constant is very small (10^{-14}). The amount of H^+ and OH^- present are 10^{-7} g-ions per lit. pH is defined as the logarithm of the reciprocal of the concentration of free hydrogen ion (H^+).

In neutral waters H^+ = OH (10^{-7}) and the pH is 7; When more H^+ activity in an acid reaction increases. The power from neutrality (10^{-7} of pH^7), say to 10^{-5}, pH is 5; when the reaction is more alkaline H^+ decreases, say to 10^{-9}, pH is 9.

Alkalinity of Natural Waters

This refers to the quantity and kinds of compounds present which collectively shift the pH to the alkaline side. In freshwaters alkalinity is mainly due to presence of bicarbonates, carbonates and hydroxides. The CO_2 - HCO_3 - CO_3^{2-} Equilibrium system is the major buffering mechanism in freshwater.

Total Alkalinity: The milliequivalents of acid necessary to neutralize the OH^-, HCO_3^- and CO_3^{2-} ions in a litre of water determines the total alkalinity.

Acidity of natural waters: Due to total available acid and the concentration of hydrogen ions.

Classification of lakes: [ON$_2$ basis of bound CO_2]:

a. Soft-water lakes - those whose waters hold small amounts of Ca & Mg salts in solution and in which the average bound CO_2 does not exceed 5 ml, l^{-1}

b. Medium-class lakes - those whose bound CO_2 falls between 5 & 22 ml. l^{-1}

c. Hard-water lakes - those which contain more than 22ml. l^{-1} of bound CO_2. The maximum may be about 50 ml. l^{-1}

CHEMICAL PROPERTIES OF WATER

Discussed in text.

ANOMALOUS CURVE

This may result from the settling of cooled. High O_2 surface water or by layering at intermediate depths of inflow waters that have a different O_2 concentration than the lake in question.

Oxygen deficit: Oxygen deficit in a lake is equal to the amount of O_2 necessary to reach saturation minus the amount of O_2 actually present.

Diel changes in O_2 Content: Diel means 24 hours: day and night; whereas **Nocturnal** is during night only and **Diurnal** means only the lighted period of the day.

A highly productive lake may be below saturation to super saturation in the course of 24 hours. even the O_2 may get nearly depleted during early morning hours.

Seasonal Changes in O_2 content

Action of O_2 on iron: Oxygen combines readily with other elements to form oxides. If sufficient soluble Fe is present then ferric oxide (orange-ochre color) is formed. The change in oxidation state of metal ions and some nutrient compounds is defined by the redox or oxidation - reduction potential.

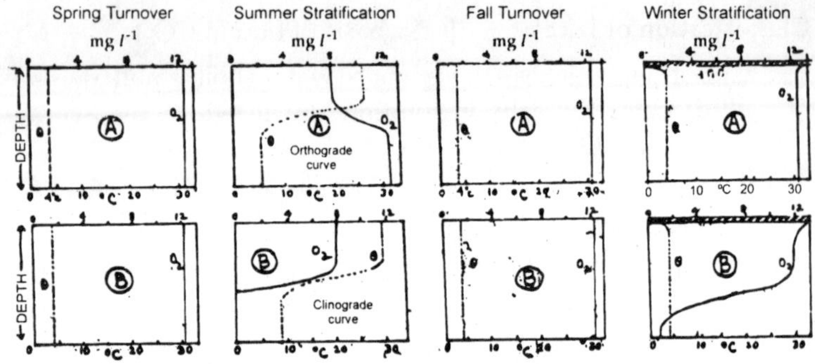

Fig. 31: Idealized vertical distribution of O_2 Conc and temperature during the 4 seasonal phases of an (a) Oligotrophic lake, (b) Eutrophic lake.

Charles law: Pressure remaining constant, volume of a given mass of a gas is directly proportional to the absolute temperature.

$$\frac{V_1}{V_2} = \frac{T_1}{T_2}$$

Boyle's law: Temperature remaining constant volume of a given mass of a gas is inversely proportional to pressure to which it is subjected

$$\frac{P_1}{P_2} = \frac{V_1}{V_2}$$

Henry's law: Temperature remaining constant solubility of a gas in a liquid is directly proportional to the pressure to which it is subjected.

Dalton's Law: Pressure exerted by a mixture or gases 'A' & 'B' = pressure exerted by 'A' + pressure exerted by 'B'.

What is pH?

Definition : It is the negative logarithm of the hydrogen ion concentration (H^+) Expressed in gram equivalents.

pH of pure water = - log (10^{-7}) = 7 (neutral)

pH of Acidic water = less than 7

pH of alkaline water = more than 7

Effect of pH on pond fish

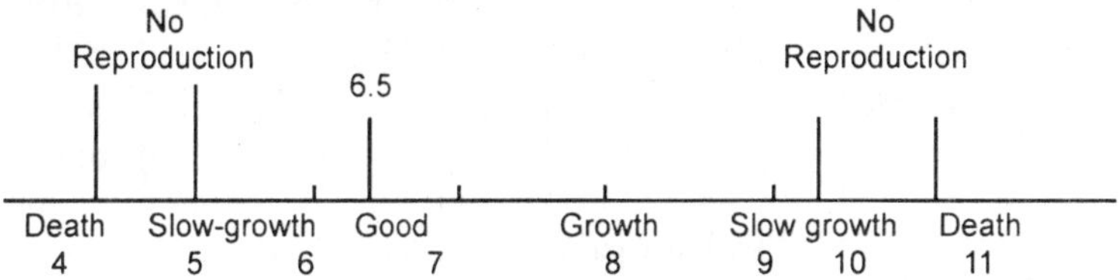

FISH GROWS BEST IN SLIGHTLY ALKALINE WATERS

Water Movement

(a) Rhythmic (waves), (b) Arrhythmic (currents)

Wind is the primary force moving lake water at all depths.

a. **Waves** Consist of rise and fall of water particles involving some oscillations but no net flow:

i. Travelling surface waves (wind-driven surface - gravity waves) - Normal waves - limited only to uppermost strata.

ii. Standing waves - standing waves cause much grater displacements of water particles.

Seiche: Seiche is an example of standing wave and consists of a local periodic rise and fall of the water level.

Surface seiches - These are the regular, resonating waves which pass from one edge to the other and back. They occur in the boundary between air and water only.

Internal seiches: Having large amplitude but short wave length occur in the thermocline. Wind induces internal seiches.

b. **Currents:** Consist of net unidirectional flows of water.

Water currents at lake surface are called **Surface drifts** water currents flow at 45° (approx) to the direction of prevailing wind (due-earth's rotation - "Coriolis effect"). Water currents below the surface move at progressively greater angles to the wind with increasing depth; finally the deepest currents flow opposite to the direction of wind. Such spiral - stair - case like flow is called **Ekman spiral**.

Lake currents range from 0->30 cm S^1

Laminar flow: Is the smooth slipping of water particles past each other or an obstruction and has little drag on moving objects.

Turbulent flow: Is the random, chaotic tumbling of the water particles around each other or any object passing through the water. It causes diffusion and spreading of a patch.

Eddies: The tumbling motions of turbulent flow are termed eddies.

When wind flows at the surface of a lake, friction develops, resulting in a shear stress at the wind water interface.

Shear instability or mixing occurs when the velocity difference produced by the turbulence flow across a density interface (Epi and hypolimnion) is exceeded and vortices formed (Fig. 32) Breaking waves contribute vertical mixing.

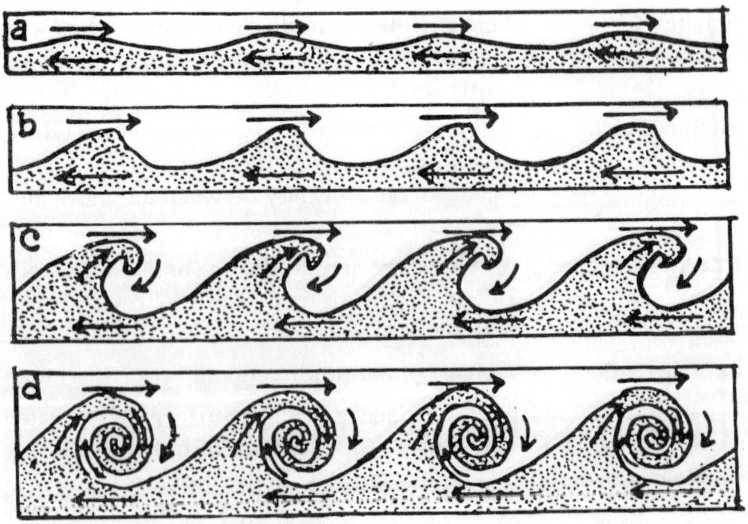

Fig. 32 : Stages in the Vortex Formation During Shear Instability on the Interface of a Stratified Two-layer system

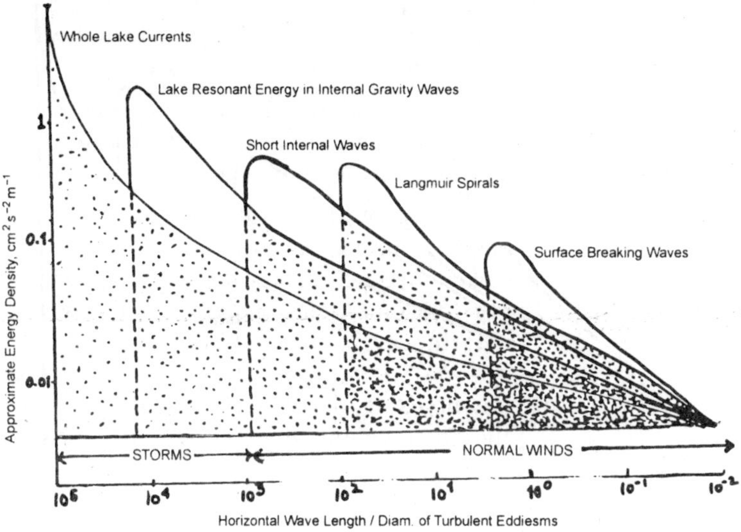

Fig. 33. Diagrammatic representation of wind-energy distribution in a hypothetical large lake.

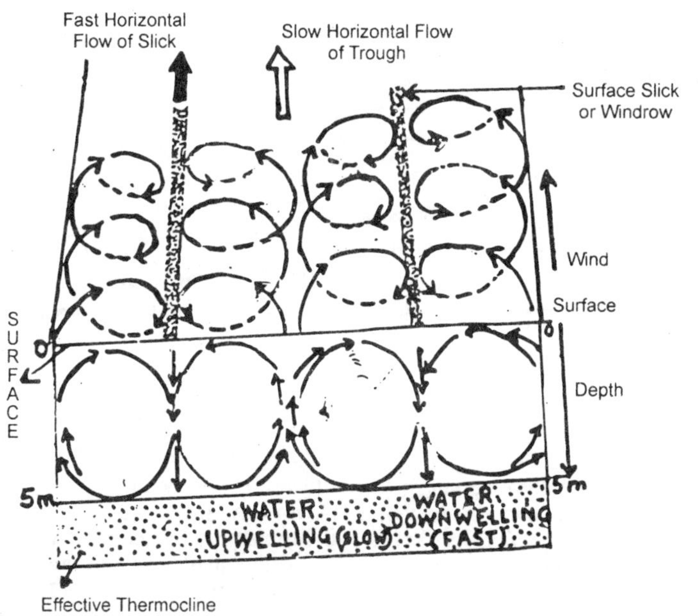

Fig. 34 : As viewed downwind, wind-produced langmuir spirals show surface water mixed down between two adjacent left - and right-handed spirals.

Evaporative cooling at lake surface, also is a major vertical mixing force; also

Convective mixing, in large exposed lakes upwelling and down welling produced by storms are a major mixing force.

Langmuir Spirals : These are produced due to an interaction between surface waves and wind-driven drift currents. (Fig. 33) These are series of adjacent vertical clockwise or counter clockwise rotating cells of water. They produce alternated areas of upwelling and downwelling, and foam accumulated above the downwelling zone, known as 'Slicks or Windrows'. (Fig. 34)

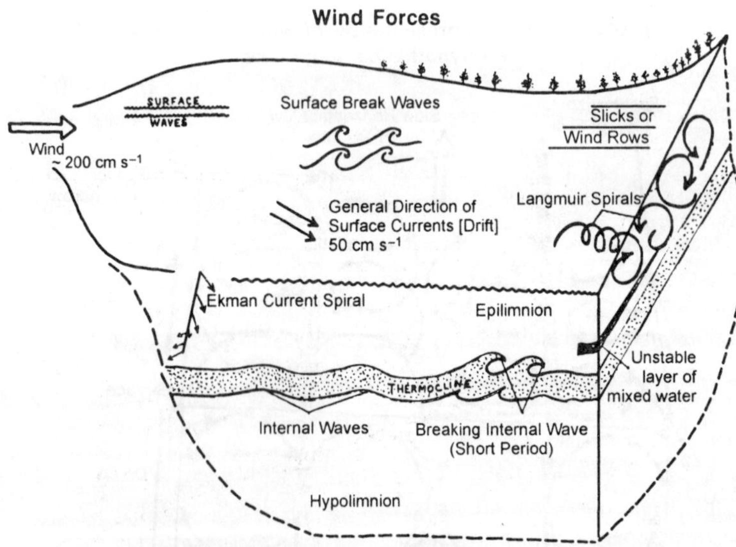

Fig. 35: Composite diagram of forces [wind, gravity, evaporation earth's rotation] and the resultant water currents and waves. Wind moves the water, gravity makes horizontal flow easier than vertical, evaporation cools surface water, which then sinks and the earth's rotation moves surface flows to the right [Northern hemisphere] to the left [Southern hemisphere]

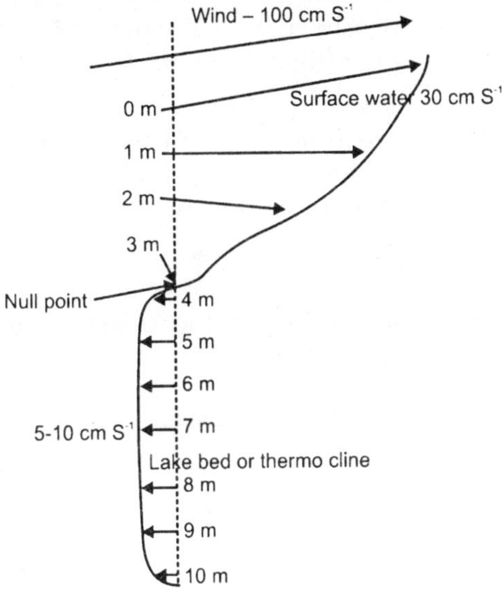

Fig. 36 : An idealized diagram of ekman spiral

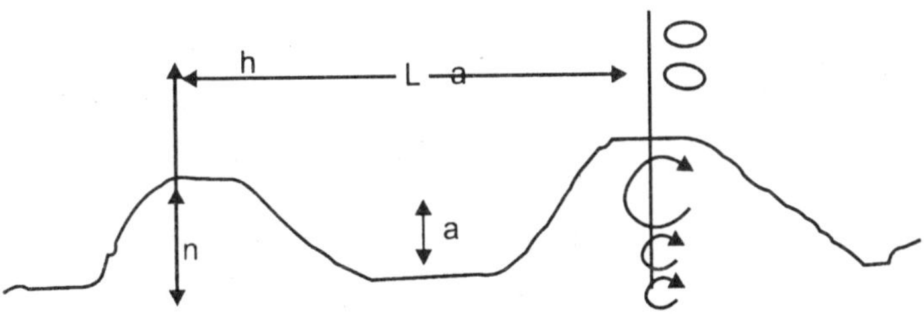

Fig. 37: Diagram of the motion of water parcels and the various liner definitions of a Rhythmic wave. The water parcels oscillate elliptically but show no net lateral motion. The amount of movement decreases exponentially with depth.
L=wavelength h=wave height a=wave amplitude (=1/2 h)

WATER AS AN ENVIRONMENT

Water has two-fold effect on living organisms in lakes:

a. Through its physical properties - medium in which organisms live

b. Through its chemical properties - supply nutrients for primary production.

Physical properties: Specific heat, latent heat, specific gravity or density, pressure, compressibility, buoyancy, viscosity and surface tension.

Density: Water is 775 times denser than air. Aquatic organisms do not have to carry their body-weight - thus save energy, also reduction of supporting tissues.

Greater density exerts a marked buoyancy on organisms. Greater the density, greater the buoyant force. The density increases with decrease of temperature and reaches maximum at 4°C after which it decreases. Sp. gravity of pure ice at 0°C is 0.9168.

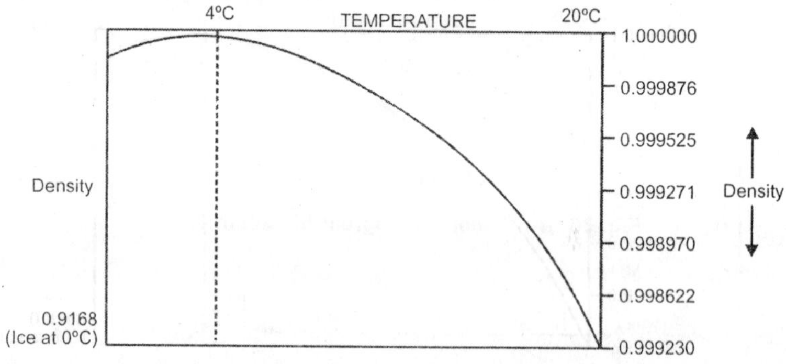

Fig. 38 : Graph showing relation between density and temperature of pure water

Table 11. Changes in the density of water with salt content.

Salt Content (per thousand)	*Density (at 4°C)*
(ean or sea ater)	

Hydrostatic pressure can lower the temperature of maximum density. The temperature of maximum density decreases about 0.1°C per 100m or depth.

(10 atmospheric pressure)

PROPERTIES OF WATER

Water exhibits some unique properties which derive from the structure of the water molecule and its tendencies to form aggregates.

Water molecules are strong dipoles and experience a strong attraction for each other and associate to form hyper-molecular linear, areal and spherical clusters. This results in a complex electrostatic system of hydrogen bridge [bonding] formation. Each water molecule is connected to four adjoining molecules by mens of hydrogen bridges forming roughly a tetrahedral arrangement. This makes water a most active chemical compound of extensive significance in the metabolism of living organisms.

Hydrogen bridge (bond) formation.

STORY OF WATER

What is the most common thing in the World?

What is the most unusual chemical in the World?

What is the most essential thing for life?

The only answer is **Water** water is life

Water is absolutely essential to every form of life

Water covers over 70% of Earth's surface

You and I are also 70% water.....do you believe?

Water is a liquid, a solid (ice) and a gas (vapour)

Unusual Properties of Water:

a. **Expansion and contraction**

When any liquid cool. - It contracts - becomes denser when water cools below 4°C. It starts expanding again so, at 0°C (ice), it is less dense. Ice is only 9/10 as heavy as same volume of water. It expands on freezing.

How it helps life? It prevents entire water column in a lake or ocean to get frozen. Surface water gets frozen while denser water at 4°C sinks down and sustains animal life.

b. **Specific Heat:** Specific heat of water is very high. It gains/loses large amount of heat without much change of water temperature. This helps in regulating earth's temperature.

c. **Boiling and melting point:** Unusual wide range of temperature between the melting point and boiling point helps life to sustain in wide fluctuations in water temperature range in NH_3 - 26.6°C; IN H_2S-7°C; IN CH_4- 3.2°C while in water 100°C.

d. **Water is a universal solvent:** Thousands of chemicals are dissolved in H_2O of the oceans. Scientists say: "it is no accident that life started in this great rich chemical soup.

   Water totally dominates the chemical composition of all organisms.

FLOATING MECHANISMS AND WATER TURBULENCE

1. **Water turbulence:** (Critical in the basic regulation of algal periodicity and production).

   Importance of water movements in the transport of particulate organic matter is well recognized. These are:

   a. Physical movements of algae into or out of the Photic zone

   b. Vertical transport of mineralized matter from lower depths and littoral regions to the open water.

2. **Floating mechanisms:** (means to remain within the photic zone) Adaptation to improve floatation or reduce sinking rates: Density of most FW Plankton which is 1.01 to 1.03 times of water, causes them to sink. Stoke's law "Spherical particles of a mean diameter of not more than 0.5 mm fall."

Various characteristics to improve flotation

a. Increasing surface - to - volume rations.

   i. Elongation into cylindrical or discoid shapes

   ii. Projection of spines or other cellular processes.

b. Production of mucilage [density of mucilage is less than the cell.

c. Gas vacuoles: These are gas filled structures in protoplasm of living cell.

d. Accumulation of fat.

C. **Inorganic Nutrient factors:** Importance of phosporus, nitrogen and other inorganic macro-and micro-nutrients for plant growth is well-known.

Liebig's law of minimum: Each organism requires a certain number of food materials and each of these materials must be present in a certain quantity. If one of these food substances is absent, the organism dies; if not absent but present in minimal quantity, the growth will be minimal. This happens even though ample amount of other required substances are present.

According to this law, "yield of any organism is determined by the quantity of that particular substance that in relation to the needs of the organism is lest abundant in the environment."

SALINITY OF INLAND WATERS

These were discussed in text.

BELIEVE IT OR NOT

Running water culture can do miracles

Table 12 : Here are a few fantastic production figures.

Sl. No.	*Country*	*Species*	*Estimated production (mt./ha/yr)*	*Source*	*Remarks*
1	Japan	*Cyprinus carpio*	1,500 (9m)	Kawamoto (1957)	Water flow (360 l/sec)
2	Japan	*C. carpio*	4,180	Hickling (1962)	Water flow (360 l/sec)
3	Japan	*C. carpio*	2,200	Chiba (1965)	Flow through From River
4	S. Vietnam	*C. carpio*	3,000	Tal (1974)	Cage culture (Mekong River)
5	Indonesia	*C. carpio*	408	Djajadiraja (1977)	Irrigation Canal
6	Kampuchea	*Pangasius sutchi* *P. larnaudii*	1,646	Hickling (1972)	Cage Culture (Mekong River)
7	USA	*Salmo gairdneri* (rainbow trout)	2,000	Tal (1974)	Raceways
8	China (Zhejiang)	Polyculture of carps	225	Coche	Cage Culture (In river : experimental)

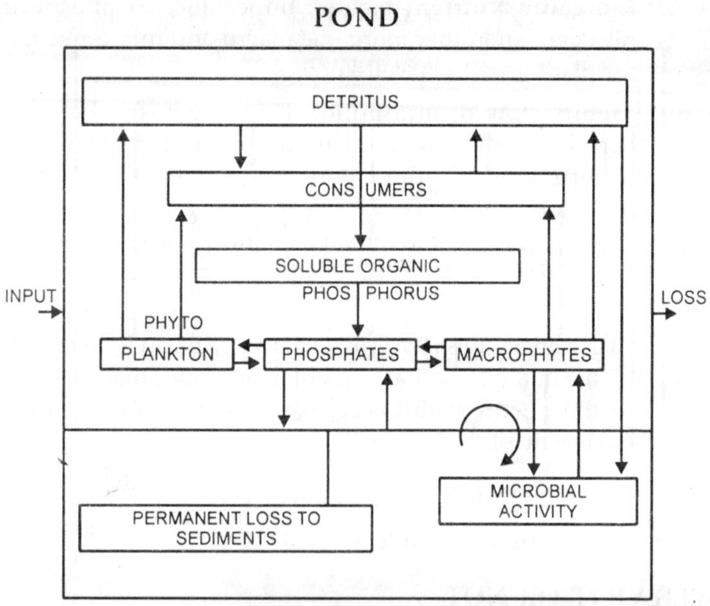

Fig. 39 : A qualitative model of the phosphorus cycle in a fishpond.

ICE: ICEBERG: GLACIERS

One-tenth of land area of earth is covered with ice. Ice melts at 0°C when mixed with chemicals melting point is lowered. Also under additional pressure ice melts at lower temperatures.

Salt: Ice :: 1 : 8 melts at -7°C. Seawater freezes at - 2.2°C. Ice floats with 1/7 to 1/8 of its volume above seawater. Ice also flows when very thick [60m - 90m].

What is an iceberg?

Icebergs are mountains of freshwater ice floating in the ocean. They are huge chunks broken off from the great masses of land ice called glaciers. Icebergs are from about 100m to several km long. Some float 3m above water surface and may be as high as 40-120m.

What is a glacier?

A glacier is a mass of ice so thick that it actually moves by the force of gravity and the pressure of its own weight. They move very slowly. A few cm to 16m a day. (slow-moving, thick masses of ice are called glaciers).

Glaciers are of two types

a. Valley glaciers (or alpine glaciers): Look like rivers of ice - mostly less than 2 km [mx: 30-80 km]

b. Continental glaciers are ice sheets, several thousand thick and cover vast land areas: Ice sheets of arctic, green land and antarctica (millions of sq. km. area reaching thickness: 3-4 km).

WHAT IS LIGHT?

These were discussed in text.

LIGHT IS AN ENERGY

Almost all energy that controls the metabolism of lakes is derived directly from the solar energy utilized in photosynthesis.

Solar energy that reaches the surface water of a lake determines the intensity and quality of light but also affects the temperature of water through radiation, evaporation and conduction.

Solar radiation incident on the lake surface is both direct sunlight and in director diffuse light which is scattered and reflected off mountains, clouds and atmospheric particles.

A beam of light falling on a water surface at a certain angle (angle of incidence), is in part penetrates the water and in part reflected. The proportion of reflection depends on the angle of incidence.

Light penetrating water is first refracted (water is a transparent liquid), then either absorbed, scattered or transmitted. Light that is absorbed is changed to heat and the intensity of light decreases. The amount of radiation retained due to scattering and absorption is termed extinction.

Factors influencing light penetration

a. Intensity at surface

b. Angle of incidence of light

c. Difference in latitude

d. Seasonal differences.

e. Diurnal differences

f. Dissolved materials

g. Suspended materials.

The amount of light that is reflected from lake surface on a clear summer day is 5-6% (average). This value increases in winter to 10%. The reflection also increases due to wave action.

The percentage of reflected light is called albedo. Scattering of light from water results in the loss of large amounts of light from lake.

Light exerts a very profound influence upon a whole series of biological phenomena in water.

HYDROGEN - ION CONCENTRATION

An important means of understanding the chemical condition of water is through the measurement of H-ion concentration.

Ion : An atom or group of atoms which carry an electrical charge.

or

The molecules of an electrolyte largely break up when in solution, into their constituent radicals, each radical carrying an electrical charge - positive or negative. These charged radicals are ions.

Electrolyte : A compound whose water solution conducts electric current.

Ionization : It is a form of dissociation of a compound in solution into ions.

Strong electrolytes (strong acid of bases) dissociate more completely than weak electrolytes.

Pure water is both an extremely weak acid and an extremely weak base, dissociating into two ions $H^+ + OH^-$ to an exceedingly small extent.

A base is any substance furnishing on ionization, the negatively charged OH^- ions.

An Acid is any substance furnishing on ionization, the positively changed H^+ ions.

A salt is any substance furnishing on ionization a positive ion other than H^+ and a negative ion other than hydrozye.

H^+ ion concentration is expressed as:

1. The no. of moles of ionized H^+ per litre
2. By pl^+ scale

pH = 1/10g 10 H^+ ion conc. in gram equivalents per litre

A Na atom loses an electron [Na -e-Na^+] and becomes positively charged. The Cl atom gains an electron and acquires a negative charge. On water solution of an electrolyte contains equal No. of positive and negative charges.

OXYGEN

These were discussed in text.

REDOX POTENTIAL

In purely inorganic chemical systems of oxidation - reduction (Redox) reactions exhibit a flow of electrons between oxidized and reduced states until an equilibrium is reached. There is a tendency for the reduced phase to loss electrons and transform to an oxidized state.

e.g. $Fe^{++}_{-e} = Fe^{+++}$

The Redox potential is proportional to the equivalent free energy change per mole of electron associated with a given reduction.

In simplest term a substance is oxidized if it loses electrons and reduced if it gains electrons. Every oxidation requires a corresponding reduction to balance electrons.

Oxidation - reduction may be illustrated as:-

$Fe^{+++}+^{e-} = Fe^{++}$ [Ferric is reduced to ferrous by gaining an electron^{e-}]

Or Fe^{++} - e- = Fe+++

[Ferrous is oxidized to ferric by losing e-]

The redox potential of a solution is a measure of the proportion of oxidized to reduced substances.

Redox potential is measured by a hydrogen electrode and is called Eh.

IF^{e-} flows from soln. to electrode - reducing conditions are said to exist in the solution.

IFe^- flows from electrode to soln $\rightarrow$ oxidizing conditions are said to exist in the solution.

IRON

Discussed in text

SULFUR (S)

Discussed in text

SILICON

The text describes the properties. A Silicon cycle is shown in Fig. 40

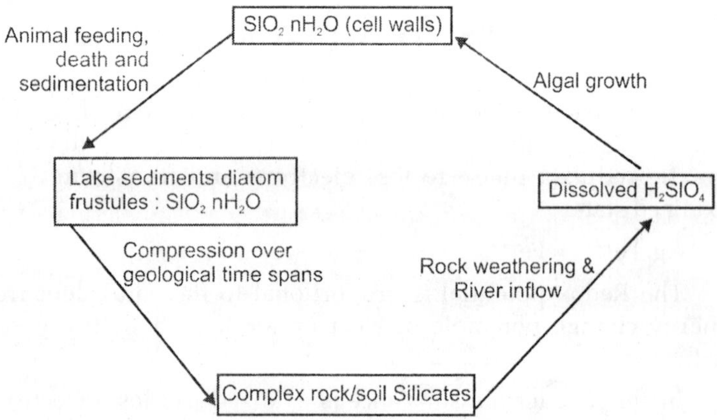

Fig. 40 : The Silica Cycle in lakes, emphasizing the role of diatoms.

PHOSPHORUS

Phosphorus is least abundant in nature and yet it limits the biological productivity of natural waters. Free P does not occur in nature.

Sources of phosphorus in inland waters:

(a) Precipitation

(b) Land run off and inflowing waters

(c) Groundwater

(d) Plant debris and animal activity

Loss of phosphorus:

(a) Out flowing water

(b) Bottom deposits

(c) Plant and animal harvest P is removed from water by all phytoplanktonic organisms.

Forms of phosphorus in natural waters :

A. Inorganic fraction

i. Soluble phosphates

ii. Sestonic acid soluble phosphate - P (mainly ferric and calcium phosphates)

B. Organic fraction

i. Dissolved and colloidal organic P (30%)

ii. Organic sestonic P (70%)

Phosphorus Cycling

Lake sediments contain much higher concentration of P than the water. The exchange of P between the sediments and overlying water is a major component of the P^- cycle in natural waters. There is a net movement of P into the sediments especially under aerobic conditions. However, in undisturbed anoxic (anaerobic) sediments P moves upward.

Bacteria are of major importance in the dynamics of P^- cycling p^- mobilizing bacteria especially of the genera Pseudomonas, Bacterium and Chromobacterium are abundant in lake sediments. The rate of release from the sediments increase markedly if the sediments are disturbed.

P is essential for plant growth although needed in a very small quantity as compared to C and N. The C:N:P ratio of plant is roughly 40C : 7N : 1P by weight.

Phosphorous : Accumulates in deposits and must be returned too the cycle by erosion, weathering, geological events and the mining activities of man. Subsequently, it is utilized by plants and animals. Bacteria plays a basic decomposing role in these events.

[From: Fundamental Concepts of Biology]

THE PHOSPHOROUS CYCLE

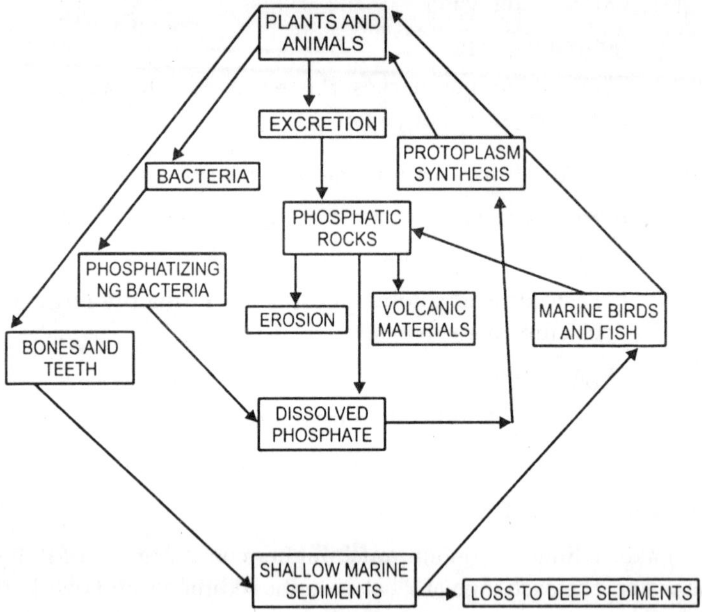

Fig. 41

The total P in lakes varies from less than 1 mg m^{-3} to even 78_9 m^{-3}.

It increases with lake productivity. Lakes rich in organic matter have higher total P concentrations. Eutrophic lakes show a considerable increase in the P content of the lower hypolimnion. This is primarily due to liberation of phosphates from sediments on reduction oligotrophic lakes show little variations in P content with depth also. In oligotrophic lakes with orthograde O_2 curve, oxidized metals such as Fe and major cations particularly Ca induce precipitation of P.

Soluble Inorganic Phosphates

Soluble orthophosphate ions are the simplest form of P in water:

$H_3 PO_4 = H^+ + H_2 PO_4^-$

$H_2 PO_4 = H^+ + HPO_4^{2-}$

$HPO_4^{2-} = H^+ + PO_4^{3-}$ [orthophosphate ion]

Presence of ionic species depends on pH.

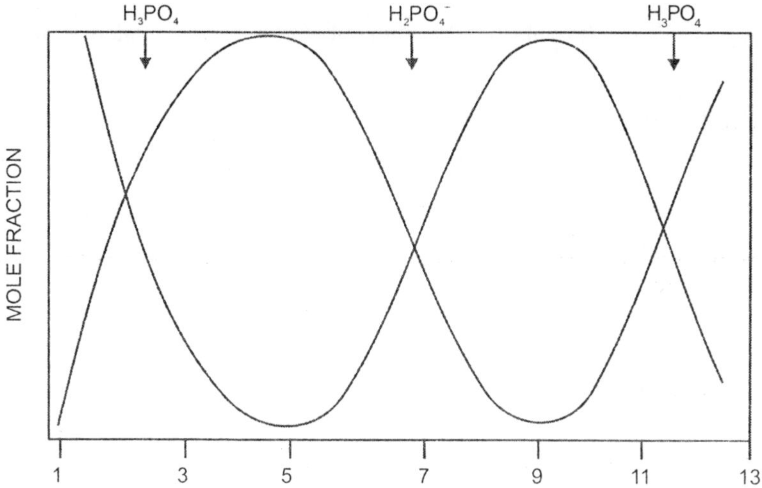

Fig. 42 : Effects of pH on the relative proportions of H_3PO_4, $H_2PO_4^-$, HPO_4^{2-} and PO_4^{3-} in an orthophosphate solution.

Inorganic soluble P is very low constituting only a few per cent of total P soluble orthophosphate, conc. are usually not more than 5-20mg l^{-1} as P and seldom exceeds 0.1mg l^{-1} as P. Even in highly eutrophic waters. The water in oligotrophic lakes often contains less than 0.001 mg of Inorganic P per litre. More that 90% total phosphorus in lakes is bound organically inorganic PO_4 or adsorbed to inorganic or to dead particulate organic materials.

CARBON DIOXIDE [CO_2]

Carbon is the primary element in organic synthesis. CO_2 provides the major source of carbon in photosynthesis and is a product of respiration by both plants and animals.

Sources of CO_2 in water:

1. By direct diffusion from air CO_2 is about 0.033% by volume in air.
2. From respiration of animals and plants
3. By decomposition of organic matter
4. From inflowing groundwater

5. In combination with other substances, mainly calcium and magnesium.

Reduction of CO_2 in natural waters:

1. Photosynthesis of aquatic plants
2. Agitation of water
3. Marl- Forming organisms
4. Evaporation
5. Rise of bubbles from depths.

Solubility of CO_2

CO_2 dissolves readily in water. The solubility of CO_2 is about 200 times that of O_2. It decreases with increase in temperature. Amount of CO_2 dissolved in water from air at sealevel is 1.1mgl^{-1} at 0°C., 0.6 mg,l^{-1} at 15°C, and 0.4mgl^{-1} at 30°C. Solubility of CO_2 increases considerably in water that contains carbonates.

CO_2 also occurs in nature in two other important forms.

(a) Bound CO_2 - As nearly insoluble monocarbonate such as Ca CO_3 or mgCO_3

(b) Half-bound CO_2 - that additional amount required to convert Ca CO_3 or Mg CO_3 into the bicarbonate, such as Ca $(HCO_3)_2$ or Mg $(HCO_3)_2$

CO_2 dissolves in water to produce carbonic acid (H_2CO_3) which dissociates into various fractions depending upon the hydrogen-ion concentration (pH).

(i) $CO_2 + H_2O = H_2CO_3$ it is a fairly weak acid and dissociated into:

(ii) $H_2CO_3 = H^+ + CO_3^-$

(iii) $HCO_3^- = H^+ + CO_3^=$

HCO_3^- is most abundant at typical pH levels of 6 to 8 [CO_2 + HCO_3^- + CO_3^{2-}] or the total inorganic carbon can be measured by titration or by modern instruments.

CARBON CYCLE

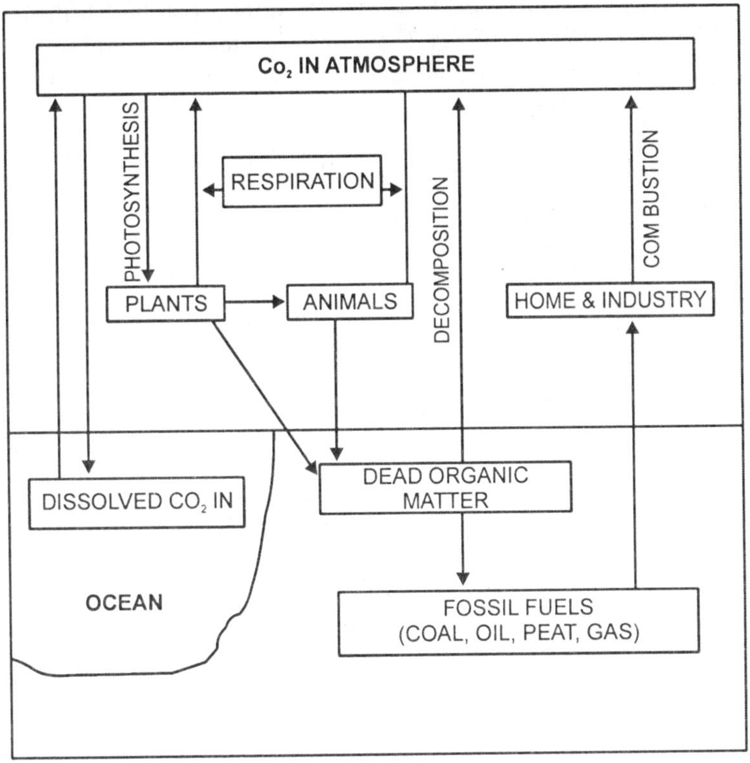

Fig. 43: Carbon cycle showing circulation of carbon as CO_2 gas between living components of environment and long term storage as fossil fuels.

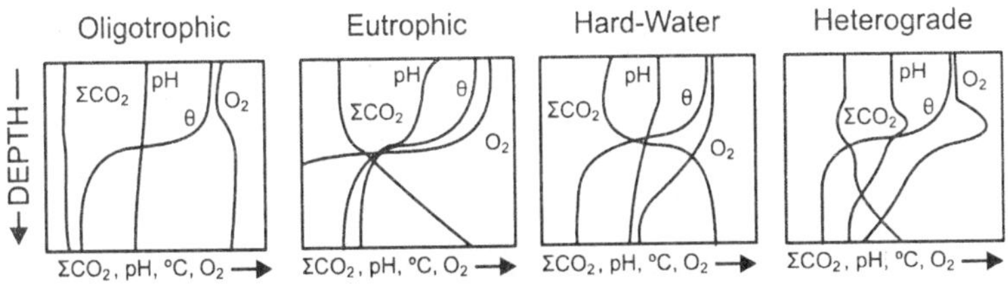

Fig. 44: Generalized vertical distribution of $\sum CO_2$, pH, °C & O_2 in stratified lakes of very low and high productivity, hard water calcareous lakes and a lake with a distinct positive hetero grade oxygen curve.

The Carbonate Buffer System: What is a Buffer?

Mixture of weak acids and their salts are called buffers. Carbonates are the most significant buffer system in natural waters and is primarily responsible for preventing rapid changes of pH in aquatic ecosystem.

Weak acid is H_2CO_3; its salts such as $CaCO_3$ and $MgCO_3$ and Ca $(HCO_3)_2$ and $Mg(HCO_3)_2$. Equation.

$H_2O + CO_2 = H_2CO_3$

$H_2CO_3 = H^+ + HCO^-_3$

$HCO_3 = H^+ + CO^{2-}_3$

A, B & C Ions present are H^+ HCO^-_3 & CO_3^{2-1}

1. If H^+ are added, it reacts with HCO_3^- to form CO_3^{2-} and are removed from the solution thus preventing a decrease in pH.

   $H^+ + HCO_3^- = H_2\ CO_3 = H_2O + CO_{2\,(A)}$

2. If OH^- is added then it reacts with HCO_3^{-1} to form CO_3^{2-} $HCO_3 + OH- = CO_3^{2-} + H_2O$ (e)

3. If CO_2 is added (due to respiration and decomposition of organic matters), CO_2 will combine with CO_3^{2-} to form HCO_3^- (J). Thus preventing drop in pH due to release of H^+

4. If CO_2 is withdrawn from water (due to photosynthesis) the HCO_3^- at once dissociates into CO_3^{-2} of $H_2\ CO_3$ - (c) thus preventing increase in pH.

   $CO_2 + H_2O + Ca\ CO_3 = Ca\ (HCO_3)_2$ (J)

   $Ca\ (HCO_3)_2 = CaCO_3 + H_2\ CO_3$ (k)

Some relationships between pH and Σ CO_2 (Free CO_2, HCO_3^- and CO_3^{2-})

Free CO_2 Dominates in water at pH 5 and below

HCO_3^- Predominates between pH 7 and 9

CO_3^{2-} is quantitatively significant about pH 9.5

Hydroxyl ions OH- are encountered in waters above pH 7.

OXYGEN PROFILE IN LAKES [Types of O_2-depth survey]

Four general types of O_2 distribution in lakes are recognized:

[a] Orthograde Curve: In less or unproductive oligotrophic lake O_2 distribution is largely a function of temperature resulting in fairly uniform orthograde curve during summer stratification. The O_2 concentration in circulating epilimnion decreases with rise of water temperature but in the meta- and hypolimnion O_2 conc. increases because of decreasing temperature and remains at saturation at the hypolimnion from just before the summer stratification.

[b] Clinograde curve: In very productive eutrophic lakes O_2 depletion occurs in the hypolimnion during summer and winter stagnation under ICE due to oxidative processes. The resulting curve in the hypolimnion anaerobic and is termed a clinograde curve. This is characterized by relatively higher O_2 content near the surface where photosynthesis is replenishes the supply.

[c] Heterograde Curve: It exhibits as irregular slope from the lake surface to the depth. It is of two types-

1. **Positive heterograde curve or metalimnetic O_2 Maxima:** In this there is an increase of O_2 in the metalimnion during stratification. As the solubility of O_2 decreases in epilimnion with increasing summer temperature and O_2 consumption in hypolimnion increases, metalimnetic O_2 maxima may occur due to photosynthetic activity of phytoplankton (O_2 producers) which concentrates at the region.
2. **Negative heterograde curve or metalimnetic O_2 minima:** This is the converse condition caused by conc. of animals when respiration dominates at some mid depth.

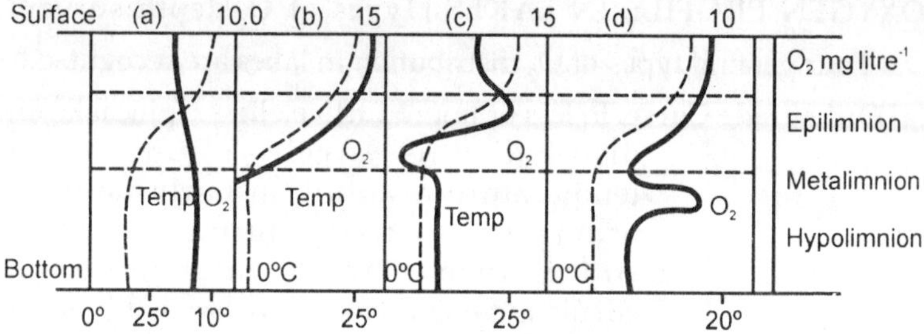

Fig. 45: Types of O_2 distribution with depth in diagrammatic form. (a) An orthrograde curve typical of an unproductive lake (b) A Clinograde curve from a productive lake (c) Positive and negative heterograde curves (d) Anomalous curves due to inflow of cool, dense O_2 rich stream.

PRODUCTIVITY

These were discussed in text.

FISH-FOOD ORGANISMS — ZOO PLANKTON

(a) Copepods

Cyclops
Canthocampus
Nauplius of Cyclops
Diaptomus
Cyclops (Female with eggs)
Nauplius of diaptomus

(b) Ostracods

Cypria
Larva of cypris
Cypris
Young ostracod

Cladocerans (Water fleas)

Daphnia Pulex
Moina brachiata
Polyphemus
Sida (dorsal view)
Diaphanosoma side
Bosmina
Alona
Moina macrocopa
Chydorus
Ceriodaphnia

Fish-Food Organisms : Zooplankton

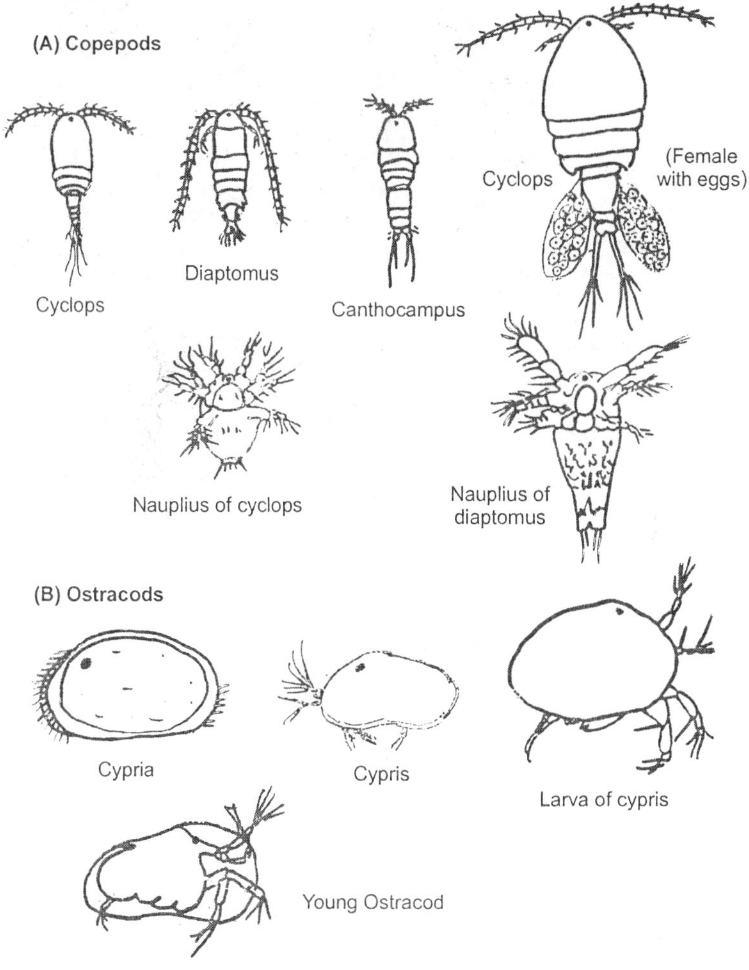

Fig. 46

Cladocerans (Water Fleas)

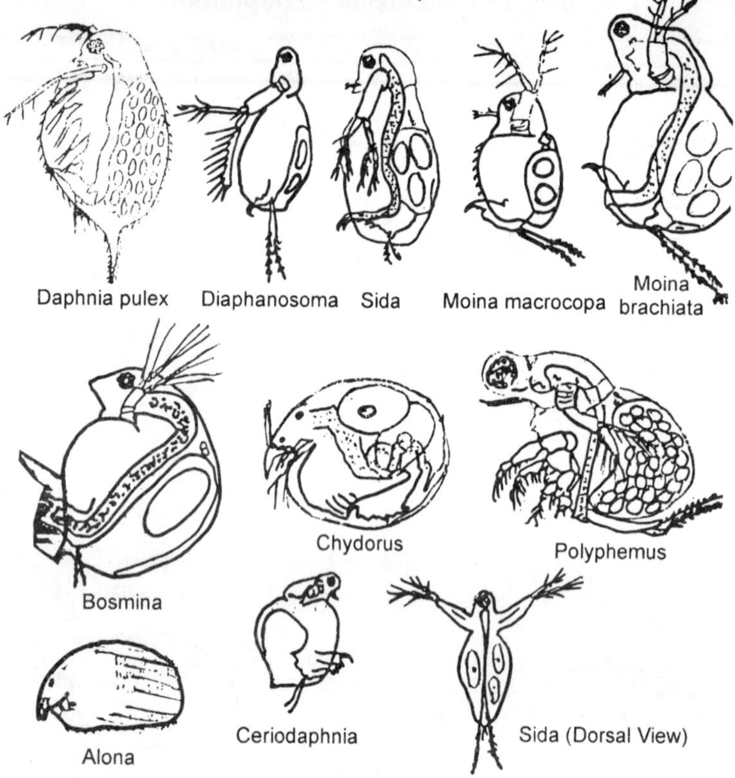

PHYLUM: ARTHROPODA

(Animals having jointed legs)

Class: Crustacea -	Hard shells; two pairs of antennae present.
Subclass: branchiopoda -	Legs are flattened and leaf-like with gills the chief respiratory organs.
Order: Anostrace -	No carapace e.g. Fairy shrimps and brine shrimps.
Order: Notostraca -	Carapace low e.g. tadpole shrimps.
Order: Conchostraca -	Carapace bivalved carapace, e.g. clam shrimps
Order: Cladocera -	Minute, usually bivalved carapace, e.g. water fleas

Subclass : Ostracoda - Minute, bivalved, compressed and enclosing entire body e.g. Cypris, Cypria.

Subclass: Copepoda - Mostly microscopic; 9 free trunk somites; no appendages in last four somites.

Order: Calanoida - e.g. Calanus (exclusively marine), Diaptomus

Order: Harpacticodia - e.g. Harpacticus

Order: Cyclopoida - e.g. Cyclops.

Branchinecta (Fairy shrimp) Apus (Tadpole shrimp)

Latonopsis (Cladocera) Lynceus (Clamshrimp)

Candona (Ostracoda) Limncaunus (Calanoida)

Macrocyclops (Cyclopoida) Bryocamptus (Harpacticoida)

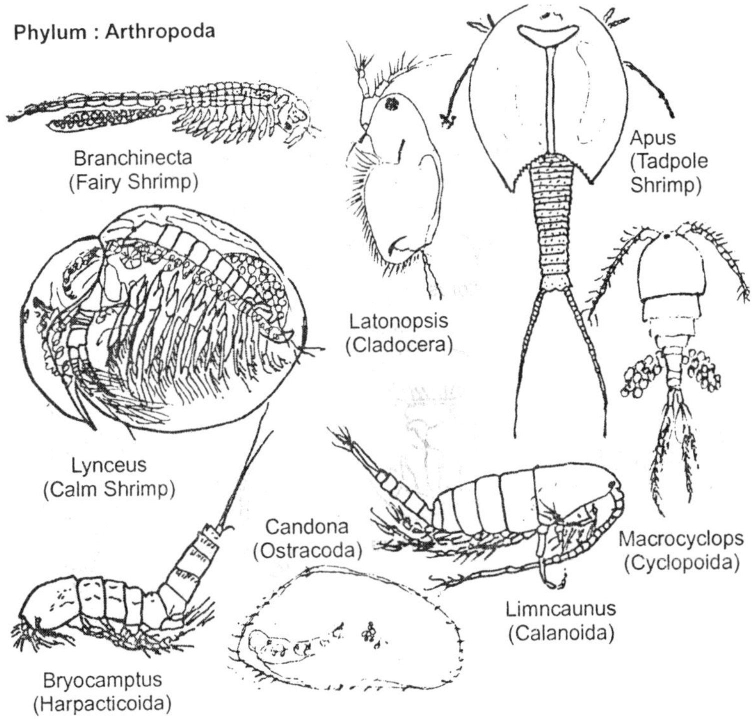

ROTIFERS (WHEEL ANIMALCULES)

Pedalia	Filinia
Tetramastix	Polyarthra
Keratella	Brachionus plicatilis
B. Calyciflorus	Brachionus
Quadridentata	Platyias
Lecane	Asplanchna
Trochosphaera	Conochilus (Colonial)
Philodina	

Rotifers (Wheel Animalcules)

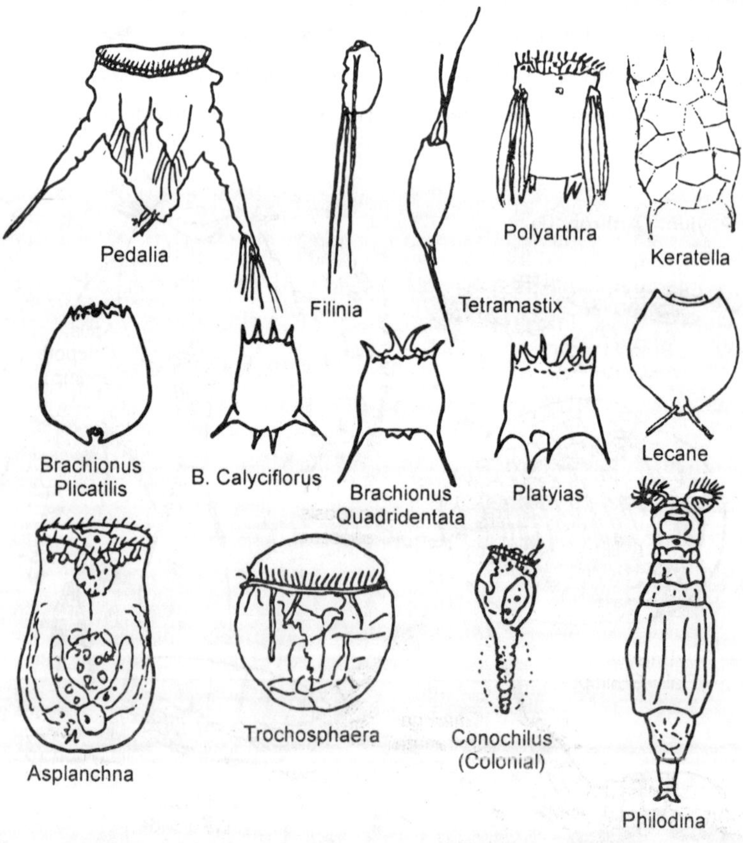

PHYLUM ROTIFERA

(Rotatoria or wheel animalcules)

Characteristics:

- Microscopic forms (40 µm - 3 mm); body distinguishable to head, trunk and foot
- Head bears a ciliated corona (Resembles wheel); mouth located in corona.
- Trunk covered with a transparent cuticle; in some thickened to form Lorica (rigid)
- Food telescopically retractable; bears 1-4 toes
- Syncytial epidermis and pseudocoel present
- Digestive system complete; pharynx cuscular (mastax), equipped with hard jaws (trophy) for sucking in and grinding food particles
- Salivary and gastric glands present

Phylum Rotifera

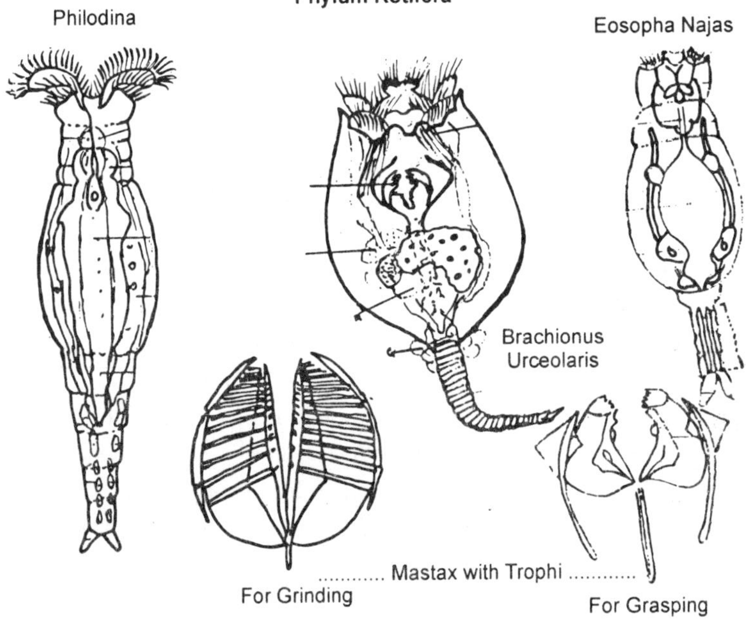

- Excretory system with a pair of protonephrial tubules with flame cells
- Nervous system with bilobed brain and sense organs.
- Reproductive system with combined ovaries and yolk glands (germovitellaria); parthenogenetic females produce diploid eggs hatching to diploid females feeding habits.

Feeding habits

- **Detritus feeders:** Feeding by means of ciliary movement of corona directing food to mouth cavity e.g. Brachionus, Keratella, Lecane.
- Carnivores (by grasping and trapping) trophy used as forceps prey and eized macerated edible parts sucred in and rest discarded e.g. Aplanchna NA.

PROTOZOA

Stentor	Paramoe cium	Lacrimaria
Stylonicnia	Vorticella	Coleps
Arcella	Amoeba	Difflugia
Centropyxis		

PHYLUM: PROTOZOA

Protozoa: Microscopic, single-celled or in colonies of similar cells, no tissues protozoans feed on detritus, bacteria, fungi, yeasts and other protozoans.

Subphylum: Plasmodroma: Organelles for locomotion are flagella, pseudopodia or none; nuclei of one kind.

Class: Sarcodina: Pseudopodia for locomotion e.g Amoeba, Difflugia, Actinophrys, Arcella.

Subphylum: Ciliophora: Cilia or sucking tentacles in at least one stage of life - history; nuclei of 2 kinds.

Class: Ciliata: Having cilia for locomotion e.g. Coleps, Stentor, Paramoecium, Vorticella, Lacrimaria, Stylonychia.

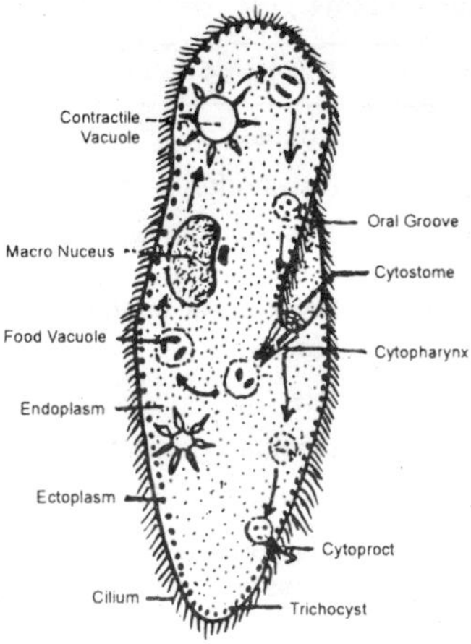

Paramoecium

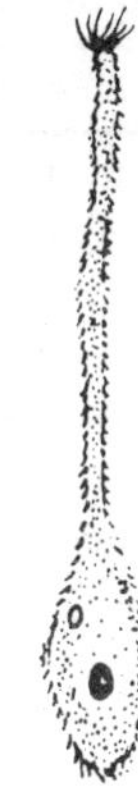
Lacrimaria

Stentor

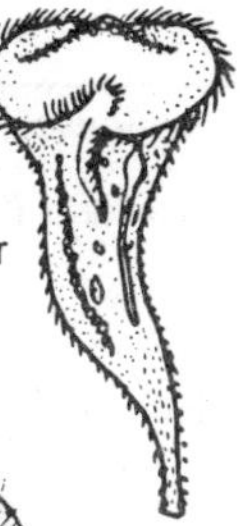

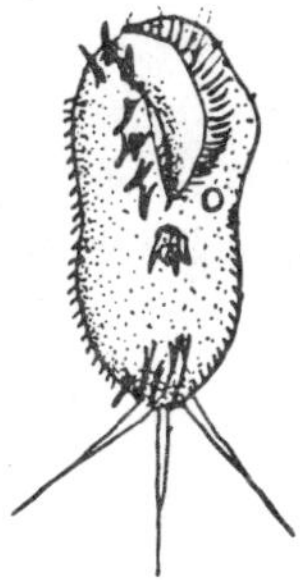
Stylonichia

Vorticella

Coleps

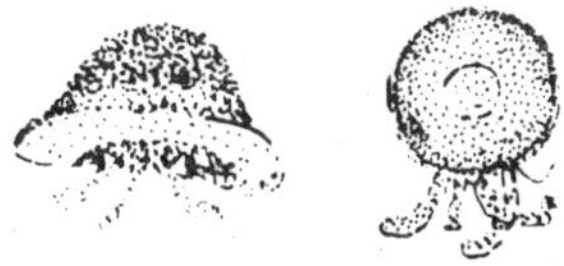
Arcella

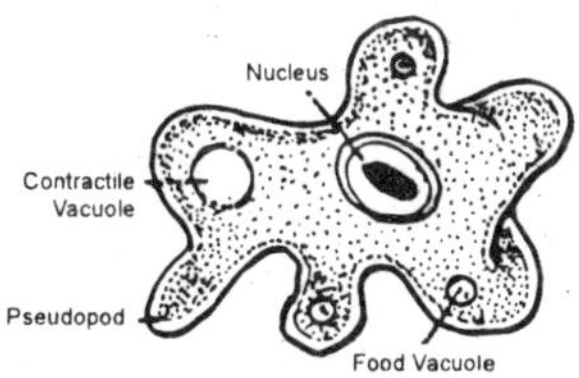

Amoeba

Difflugia

Centropyxis

Table 13 : Classification of Benthic Invertebrates Common in Lakes and Streams

Taxonomic Group	*Food/Feeding habit*	*Examples*
Turbellaria (Flatworms)	Carnivores	Dugesia
Nematoda (Roundwords)	Carnivores, Herbivores	Dolichodorus
Annelida Oligochaeta (earthworms) Hirudinea (leeches)	Sediment grazers Carnivores, Detrivores	Tubifex (Red worms) Haemopsis (Horse leech)
Molluscs	Grazer	Pila
Gastropoda (Snails)	Filter feeder	Limnaea (pond snails)
Pelecypoda (Bivalves)		Planorbis
		Anodonta (swan mussel)
Crustacea	Detrivores	Gammarus, Astacus
Malacostraca (cray fish amphipods,		
Insecta	Carnivores	Sialis
Megoloptera (dobsonfils)		
Plecompera (stoneflies)	Mostly carnivores	Nemoura
Odonata (dragon flies)	Raptorial carnivores	Libellula, Anax
Ephemeroptera (mayflies)	Mostly scrapers, grazers	Baetis, ephemerella
Hemiptera (water bugs)	Beaked Carnivores, Herbivores	Notonecta (Backswimmers)
Trichoptera (caddis flies)	Mostly filter feeders scrapers	limnephilus, hydropsyche
Coleoptera (beetles)	Raptorial carnivores	Dytiscus, cybister (diving beetles)
Diptera (two winged flies)	Filter feeders raptorial	Culex (mosquito)
Family: Culicidae	carnivores	Chaoborus (Corethba) (Phantom midge)
Family: Simulidae	Filter feeder	Simulus (black fly)
Family: Tendipedidae (chironomidae)	Filter feeder, scraper carnivore	Tendipes (Chironomus) (bloodworms, true midge)
		Calopsectra (Tanytarsus) (Midge)

Chaoborus (phantom larva)

Bloodworm

Tendipes (chironomus) Tubifex

PHENOMENON OF CYCLOMORPHOSIS

Cyclomorphosis in Rotifers

Seasonal polymorphism: Some part of the organism changes markedly :

(a) Elongation in relation to body width: Mid-summer population can be 5 times as long as wide; nearly spherical in spring e.g. Asplanchna.

(b) Enlargement with the formation of body wall outgrowths of humps: This is caused by tocopherol content of plant-derived food e.g. *Asplanchna sieboldi.*

(c) Reduction of size at higher temperature in summer: reduction of spine length increases sinking speed in summer in Keratella.

(d) Production of lateral spine: posterior spine elongates in brachonus calyciflorus in presence of the predator Asplanchna.

Cyclomorphosis in Cladocera:

Cyclomorphosis in cladocera occurs more conspicuously than any other group. In Daphnia it occurs with:

(a) Increasing water temperature [in spring]: A gradual extension of anterior part of head to form a crest or helmet also there is marked decrease in no. of parthenogenetic eggs per individual.

(b) Water turbulence - incrased antennal beating rate in order to maintain vertical orientation to light.

Seasonal cyclomorphosis in Ceratium:

In mid-summer when water gets warms and less viscous, Ceratium reduces the rate of sinking out of the photic zone by the reduction of cell dimensions, increase divergence of the horn and by the development of the 4th horn.

HUMAN POPULATION AND ENVIRONMENTAL EXPLOITATION

Many of the current ecological problems of the environment are really based on rather simple ecological principles. The balance of nature remains a balance only if there exists a proper give and take relationship among the following four groups:

- The producers that is the autotrophs or photosynthesizing organisms that capture solar energy and use to produce food stuffs, directly or indirectly, for all forms of life.
- The consumers that is the heterotrophs or organisms living either directly or indirectly off food products provided by the autotrophs.
- The decomposers that is the organisms, of decay, primarily bacteria but also fungi, and
- The inorganic world that is, the nitrates, phosphates carbonates etc. all of which are needed in balanced amounts as nutritional supplements by both producers and consumers.

One thing must be made perfectly clear of all the living organisms on earth, only one-man has ever seriously threatened the delicate natural balance existing among these four groups. He is now increasing his threats to the point that threatens not only the existence of non-human organisms but also his own, for man is the only organism that pollutes. He has brought about this pollution by two main means, and both are intimately connected to the fact that man is virtually the only organism that manipulates his environmental to serve his own needs and desires.

Reference Books of Limnology

1. Alabaster, J. S. (Ed.) : 1977. *Biological Monitoring of Inland Fisheries.* Applied Science Publ. London, 226 p.
2. Allee, W. C. *et al.* 1949 : *Principles in Animal Ecoogy,* Philadelphia, 837 p.
3. Cole, G. A. 1968 : *Textbook of Limnology* (2nd ed.) Mosby, S. Louis, 426 p.
4. Conlu, P. V. 1982. *A Guide to Philippine Flora and Fauna* vol. 1 : Endemic, rare and economically important fishes. 129 p.
5. Frey, D. G. 1963 : *Limnology in North America.* University of Wisconsin Press. Madison. 734 p.
6. Goldman, C. R. 1968 : Aquatic Primary production. *Am. Zool.* 8 : 31-42, 1968.
7. Goldman, C. R. & A. J. Horne 1983 : *Limnology.* McGraw-Hill Co., N. Y., 464 p.
8. Golterman, H. L. 1975 : *Physiological Limnology.* Elsevier Scientific Publc. Co., N. Y., 489 p.
9. Hutchinson, G.E. : *A Treatise in Limnology* vol. 1 (1957) : *Geography, Physics and Chemistry,* 1015 p; vol. 2 (1967) : *Introduction to Lake Biology and the Limnoplankton,* 1115 p; vol 3 (1975) : *Limnological Botany,* 660 p. Wiley N. Y.
10. Lagler, K. F. 1949. *Studies in Freshwater Fishery Biology,* Ann Arbor, Mich. 240 p.
11. Lind, O. T. 1974 : *Limnology,* C. V. Mosby Co. London.
12. Mamaril, A. 1978 : *A Guide to Philippine Flora and Fauna* Vol. 1 : Zooplankton. Mimeo 236 p.
13. Marshall, N. B. 1966 : *The Life of Fishes.* Universe Books, N. Y., 402 p.
14. Mellanby, H. 1938. *Animal Life in Freshwater.* London 296 p.
15. Needham, J. G. & J. T. Llyod 1930 : *The life of inland waters* (2nd ed.) Springfield 438 p.
16. Pantastico, J. B. 1977 : Taxonomy of the Freshwater Algae of Laguna de bay and vicinity, *Bullein No. 61 NRCP,* Philippines. 251 p.

17. Pennak, R. W. 1978. *Freshwater Invertebrates of the United States*, Wiley, 803 p.

18. Ruttner, F. 1963 : *Fundamentals of Limnology*. Toronto, U. Press, 307 p.

19. Thienemann, A. 1926. *Limnologie*. Breslau, 108 p.

20. Umali, A. F. 1950 : *Key to the families of common commercial fishes in the Philippines*. Res. Rep. 21, fish and Wildlife Service, US Dept. of Interior. 47 pp.

21. Umaly, R. C. and L. A. Cuvin. 1988 : *Limnology : Lab and Field guide - Physico-chem-bio factors*. National Book Store, Manila. 322 p.

22. Ward, H. B. & G. C. Whipple. 1959 : *Freshwater Biology*. Wiley N. Y., 1248 p.

23. Welch, P. S. 1948 : *Limnological Methods*. Philadelphia, 381 p.

24. Welch, P. S. 1952 (2nd ed.) : *Limnology*. McGraw-Hill Book Co., p. 538.

25. Wetzel, R. G. 1975 : *Limnology*, Saunders, Philadelphia 734 p.

26. Whipple, G. C. 1927. *The microscopy of drinking water* (4th ed.) N.Y., 586 p.

27. Wingberg, G. C. (Ed.) 1971 : *Methods for estimation of production of aquatic animals*. Academic Press, London, 175 p.

PART THREE

ICHTHYOLOGY

* * *

In all the Manuals of Prof. Chaudhuri's most attractive and fascinating presentation was his hundreds of coloured slides and multi-coloured transparencies. For breavity of this treatise and short budget allocation those could not be reproduced. The readers may have to refer to text books and research papers. This has minimized the flavour of the book.

1
GENERAL CONSIDERATIONS

Definition

Ichthyology is the science of study of fishes.

Greek words: *Ichthyos* = Fish and *Logos* = study.

Definition of Fish : Fishes are cold blooded aquatic vertebrates which breathe through gills and propel and balance by means of fins.

FISH is a very important animal because it occupies a position in the evolutionary line where *Nature* has made a bold attempt to develop. The presence of FISH on earth antedates all other vertebrates by more than 100 million years.

FISH as an ancestor of MAN

"Without piscine ancestry, man might never have evolved." Nature took a bold step to evolve fish slowly from water to land vertebrates; from gills to lungs and from fins to limbs of tetrapods. However, in its attempt, fish had to sacrifice tremendously. Fossil records show that many groups of fishes have failed to survive during the process and are now extinct.

Fish developed the ground plans and basic functions of the ten organ systems including such striking features as sight, internal fertilization, intra-uterine nourishment, live birth and presumably learning and memory.

Fish developed the modern idea of submarine in its swim bladder an excellent hydrostatic organ, developed electricity and luminous organs long before mans invention of electricity and light and experimented with production of sound which led to the amphibians, the first vertebrate to croak.

Systematic Position of FISH in the Animal Kingdom

Fishes belong to the Superclass: *Pisces* under the Sub-Phylum: *Gnathostomata* of the group : *Vertebrata.*

The FISH "describes a life form, rather than a taxonomic group." FISH as a group, has a unique position:

1. Fishes are the most diversified forms having adapted to diverse ecological and environmental conditions.
2. Fishes are the most numerous (over 20,000 species) among vertebrates.
3. Fishes inhabit almost all habitats (fresh-, brackish and marine) from highest elevation of 5 km (above sea level) to the deepest ocean of 11 km depth.
4. Fishes have acclimatized to temperature below freezing point at Antarctic to over 40°C in some hot springs.
5. Fish size vary from the minute pigmy goby (12 mm) of the Philippines to the largest whale shark having a maximum length of about 21 m and a weight of 25 m.t (according to Lagler *et al.*).
6. Fishes influence the structure of aquatic ecosystems by their highly diversified voracious feeding habits.
7. Fishes interact with various levels of aquatic food chains and occupy practically all levels excepting the producer.

Fishes are very important to man

a) *as a staple food and a source of cheap protein:* Fish is the main source of animal protein for the vast majority of Asians and major population of most of the developing countries;
b) *as a thing of beauty* (ornamental fish): From aesthetic point of view, people keep live fishes in their homes for the simple pleasure of observing the beauty and behaviour of fishes;
c) *as a means of recreation* (sport fishing): Millions of people are indulged all over the world in sport fishing. Bait-fish culture is a huge industry. In USA alone, the total retail value of bait fishes sold every year would be between US $ 60-80 million;
d) *as a friend to man:* Fishes are used for the biological control of the noxious pests of many dangerous human diseases, such as, malaria, filarial and several parasitic diseases

where snails act as the secondary hosts.

e) *as a valuable laboratory animals:* Fishes are very often used as useful test animals for toxicity study and several other laboratory tests;

f) *as a good and useful pet* (probably next to dog): Fishes have proved to be a good pet and get domesticated. One advantage over dogs is that while dog barks and disturbs, fishes are quiet pets.

g) *as a means of livelihood:* Fishing and aquaculture generate lucrative business and provide livelihood to millions of fishermen and fish culturists. These also include aquarium and bait-fish industries.

2
ORIGIN AND EVOLUTION OF VERTEBRATES (Geologic Time Table)

Table 1

Era	*Period*	*Epoch*	*Biological Events*	*Years before present*	
CENOZOIC	Quaternary	Recent Pleistocene	Modern man Early man	11,000 yrs. 0.5-3 m	Age of man
	Tertiary	Pliocene	Large carnivores	7 m	
		Miocene	Abundant grazing mammals	25 m	Age of
		Oligocene	Apes, monkeys, whales	40 m	Mammals
		Eocene	Radiation of placentals	60 m	and birds
		Paleocene	First placental mammals	70 m	
MESCZOIC Late	Cretaceous		Climax of giant land & marine rep-tiles, followed by extinction	(65 m) 135 m	
PALEOZOIC	Jurassic		First birds, first mammals dinosaurs abundant	(45 m) 180 m	Age of reptiles
	Triassic		First dinosaurs; mammal-like reptiles	(50 m) 230 m	

contd...

Table 1 – *contd...*

Era	*Period*	*Epoch*	*Biological Events*	*Years before present*	
	Permian		Radiation of reptiles; Displacement of amphibians	(50 m) 280 m	
Middle	Carboni-ferous	Pennsyl-vanian	First reptiles	(30 m) 310 m	Age of amphibians
		Mississipian	Radiation of amphibians, abundant sharks	(35 m) 345 m	
	Devonian		First amphibians; Freshwater fishes abundant	(60 m) 405 m	Age of Fishes
	Silurian		First jawed fishes	(20 m) 425 m	
Early	Ordovician		Ostracoderms (First vertebrates) Abundant marine invertebrates	75 m 500 m	Age of molluscs & echino-derms
	Cambrian		Origin of many invertebrate phyla & classes; trilobites dominant	(100 m)	Age of trilobites
	Precambrian period		Proterozoic... 1200 - 600 m; Archeozoic... 2000 - 1200 m; Azoicera... 3162 - 2000 m		

*M = million years (Arbitrary date)

3 CLASSIFICATION

The principle used in classifying animals involves the division, or the separation of large groups into smaller groups according, in the main, to certain distinctive traits which they possess in common.

A system of classification is a storage, retrieval, and communication mechanism for biological information. This is the science of *Taxonomy*.

Taxa	[Hierarchial system of classification]			
1. Kingdom	→	Sub Kingdom		
↓↑				
2. Phylum	→	Sub phylum		
↓↑				
3. Class	→	Sub class	→	Super class
↓↑				
4. Order	→	Sub order	→	Super order
↓↑				
5. Family	→	Sub family	→	Super family
↓↑				
6. Genus	→	Sub genus		
↓↑				
7. Species	→	↓	→	Super species
↓↑				
Individual		Sub species	→	variety

Binomial Nomenclature

Each species has a latinized name composed of two words. The first word is the genus, written with a capital initial letter; the 2nd word is the specific name that is peculiar to the species and is written with a small initial letter. The genus name is always a noun, and the species name is usually an adjective that must agree with the gender of the genus.

PHYLUM : CHORDATA

Principal Characters : Presence of:

1. Noto Chord;
2. A Hollow-dersal Nerve Tube
3. Gill Slits Connecting Pharynx, at least some time during Life.

Group A : Acrania : Lacks Cranium and Brain.

Sub Phylum: 1. Urochordata Tunicates/Ascidians

Sub Phylum: 2. Cephalochordata Amphioxos.

Group B : Vertebrata : With Brain, Cranium and Vertebral Column.

Sub Phylum: 1. Agnatha No Jaws.

Class: 1. Ostracodermi [Extinct]

Class: 2. Cyclostomata lampreys and Hagfishes

Sub Phylum: 2. Gnathostomata Mouth with Jaws

Superclass: 1. Pisces All Fishes.

Class (1) Placodermi [Extinct]

Class (2) Chondrichthyes Cartilaginous Fishes

Class (3) Osteichthyes Bony Fishes

Super Class (2) : Tetrapoda

Class (1) : Amphibia - Frogs, Toads, Salamanders.

Class (2) : Reptilia - Lizards, Snakes, Turtles.

Class (3) : Aves - Birds.

Class (4) : Mammalia - Cat Whale, Ape, Man.

Sub Phylum: 3. Hemichordata Balanoglossus

(According to earlier classification; recent classification excludes this group from Chordata)

Class : Chondrichthyes

(Sharks, Rays, Skates And Chimaeras)

Characters :

Notochord like a string of beads; Vertebrae Cartilaginous; Gill Arches Cartilaginous. Paired Fins and Paired Nostrils present.

Subclass:

1. *Elasmobranchii (Selachii)*

Characters:

Gills/Gill Clefts: 5 to 7 pairs; Spiracle present; Placoid scales present or absent; Cloaca present. Example: Sharks, Rays and Skates.

2. *Holocephali*

Characters:

Gills: 4 pairs; Gill cleft single pair; Spiracle absent; Lacking scales; Cloaca absent. Example: Chimaeras.

Class : Osteichthyes

Characters:

Notochord like a cord string of beads or separate beads; Vertebrae bony; Gill Arches bony paired fins and paired nostrils present.

Subclass:

1. *diphoni*

characters:

Paired fins with a pronounced Lobate, Fleshy, Scaled base; Excurrent Nares internal in Lung fishes not in Lobefins. Example: Lungfishes (Dipnsusti): Protopterus and Neoceratodus , Lobfins (Ceossopterygii) such as : Coelacanth (Latimeria).

2. *Actinopterygii*

Characters:

Paired fins without a pronounced lobate fleshy, scaled base; Excurrent Nares usually external e. g. Modern Bony Fishes (Teleostel). Gars and Browfins (Mosostei), Sturgeons and Paddle Fishes (Chondrostei).

Class : Cyclostomata

Characters:

Jawless, Notochord, cordlike; Main axial skeleton cartilaginous; No gill arches; Branchial basket present; Paired fins absent; Single median nostril. Example: Lampreys and Hagfishes.

Subclass:

1. Petromyzontes:

Teeth; Nasal sac not connected to mouth; Gill pouches 7 pairs. Example: Petromyzon, Lampetra.

2. Myxini:

Mouth terminal with 4 pairs of tentacles; Nasal sacs connected to mouth; Gill pouches 5-15 pairs. Example : Myxine, Bdellostoma.

How do we define FISH?

Fishes are Aquatic Vertebrates which breathe through Gills and Propel and Balance themselves by Fins.

Of the vertebrate group FISHES are the most numerous and most diversified. They have invaded all waters from Arctic and Temperate to the heat of Tropical Zones. Variety of Habitats has changed their bodies to different shapes and sizes which are quite distinct from each other. The following common names of different groups of fishes give an idea about the great diversity of their forms:-

Resembling Animals Body Parts

Alligator Fish	Bat Fish	Boar Fish	Butterfly Fish	Cow Fish	Coral Fish
Cat Fish	Dog Fish	Dragon Fish	Elephant Fish	Frog Fish	Goat Fish
Hawk Fish	Horse Fish	Lion Fish	Lizard Fish	Parrot Fish	Pig Fish
Pony Fish	Porcupine Fish	Rabbit Fish	Rat fish	Sable Fish	
Scorpion Fish	Shrimp Fish	Squirrel Fish	Snail Fish	Snipe Fish	Tiger Fish
Toad Fish	Omicord Fish	Viper Fish	Wolf Fish	Worm Fish.	

CUTTLE FISH?

Colours:

Black Fish	Blue Fish	Gray Fish	Red Fish	White Fish	Finger Fish
Head Fish	Jaw Fish	Nose Fish	Quill Fish	Tongue Fish	Bill Fish
Horn Fish	Lung Fish.				

Vocation/others:

Surgeon Fish	Sergeant Fish	Soldier Fish	Shepherd Fish	Pilot Fish	Angler Fish
Monk Fish	Cardinal Fish	Damsel Fish	Angel Fish	Lady Fish	Bishop Fish.

Plant Origin:

Leap Fish	Kelp Fish	Pine Cone Fish	Reed Fish.

Food:

Butter Fish	Milk Fish	Toe Fish.

JELLY FISH?

Gems/Metals:

Gold Fish	Sapphire Fish	Pearl Fish.

SILVER FISH?

Things:

Balloon Fish Band Fish Blowgun Fish Cigar Fish Comb Fish Cornet Fish
Cutlass Fish Drum Fish File Fish Glass Fish Drum Fish File Fish
Guitar Fish Hatchet Fish Hammer Head Knife Fish Lantern Fish Needle Fish
Oar Fish Pipe Fish Prow Fish Paddle Fish Rudder Fish Ribbon Fish
Rag Fish Razor Fish Saw Fish Spade Fish Spear Fish Spike Fish
Stock Fish Sword Fish Thread Fish Tile Fish Trumpet Fish Trunk Fish
Trigger Fish.

Shanimate Object:

Mud Fish Stone Fish Sand Fish Rock Fish.

SHELL FISH?

Celestial Bodies:

Sun Fish Moon Fish Globe Fish Paradise Fish.

STAR FISH?

Cave Fish Cling Fish Crest Fish. Flat Fish Flying Fish Gibber Fish
Klip Fish Killi Fish Lump Fish Poison Fish Sheet Fish Scald Fish
Spook Fish Surf Fish Round Fish Thindia Fish Weak Fish. also
Hag Fish Cod Fish Gar Fish.

4
External Morphology

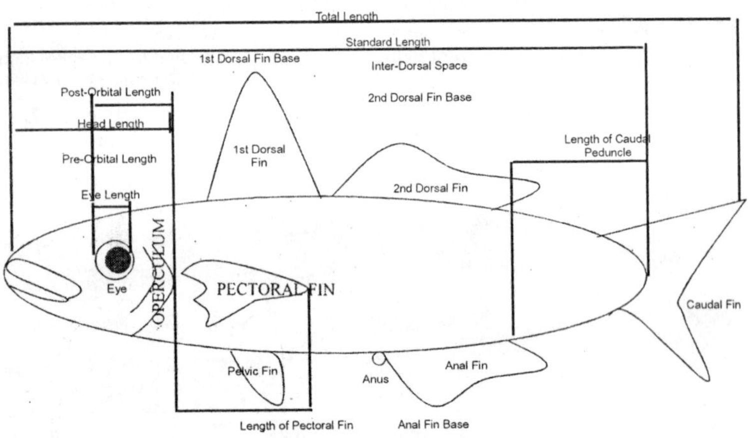

EXTERNAL MORPHOLOGY OF FISH

Mouth Types

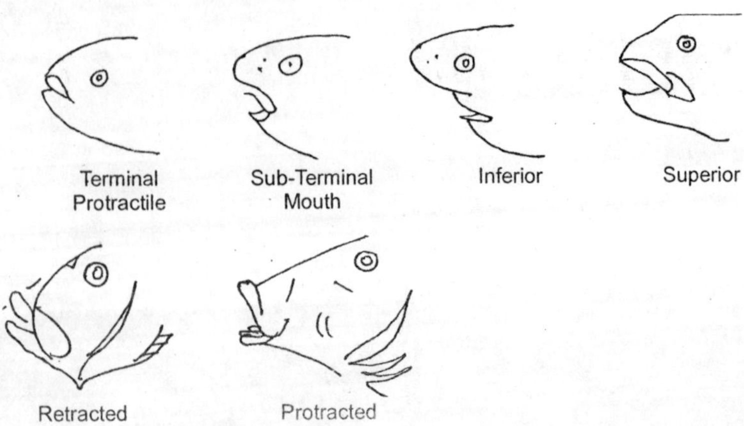

Teeth Types

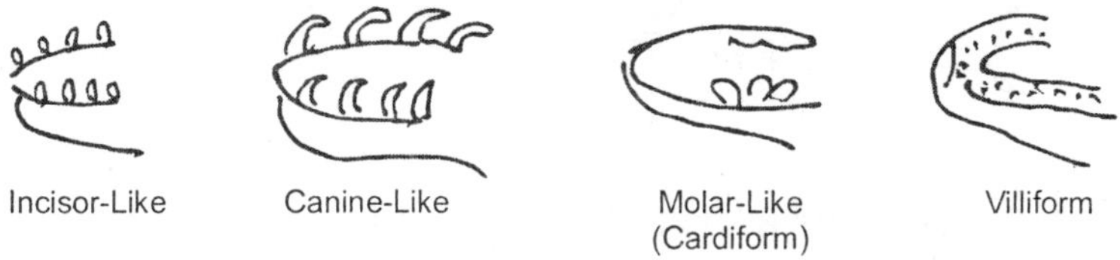

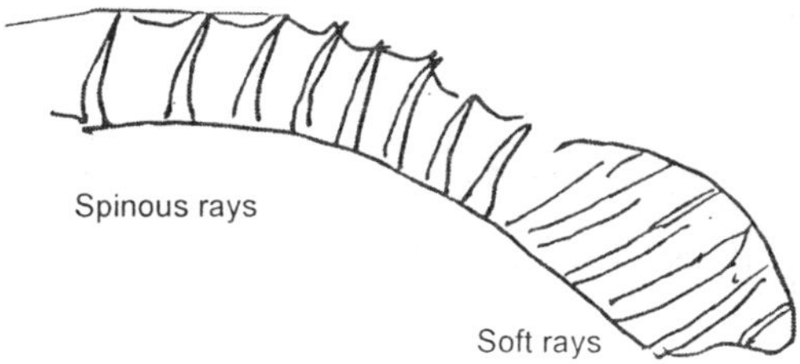

Fin Types

Scale Types

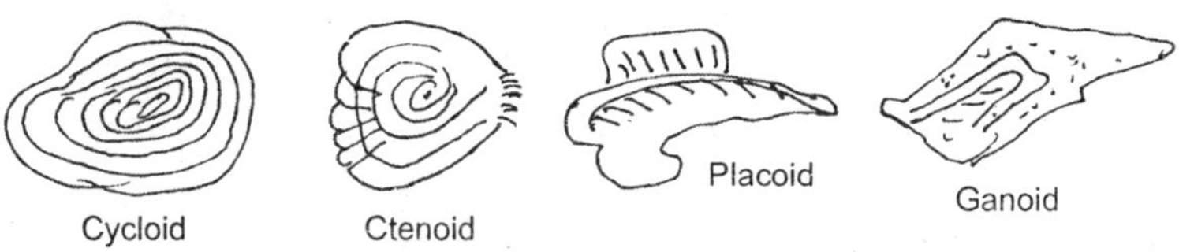

Caudal Fin Types (Homocercal)

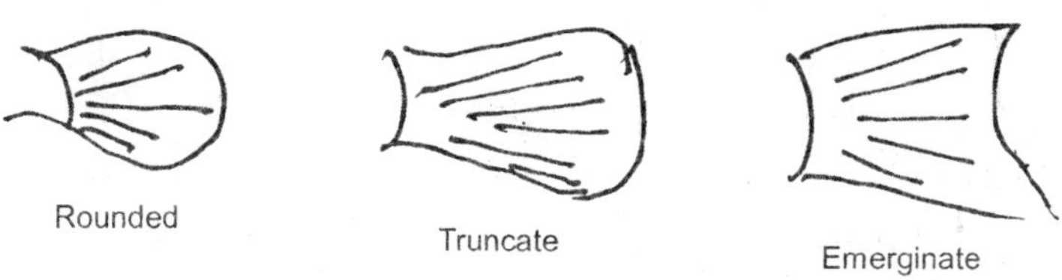

POINTED (SEPARATED FROM DORSAL & ANAL FINS)

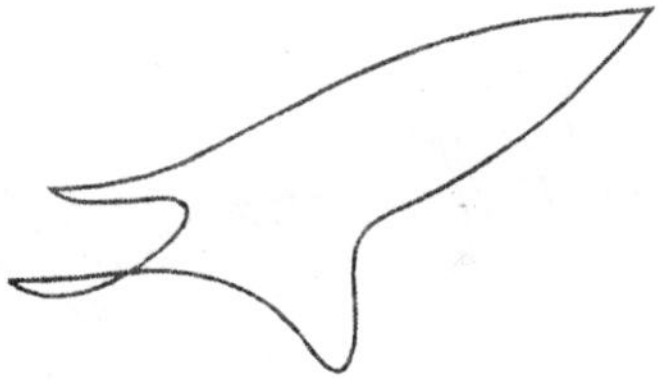

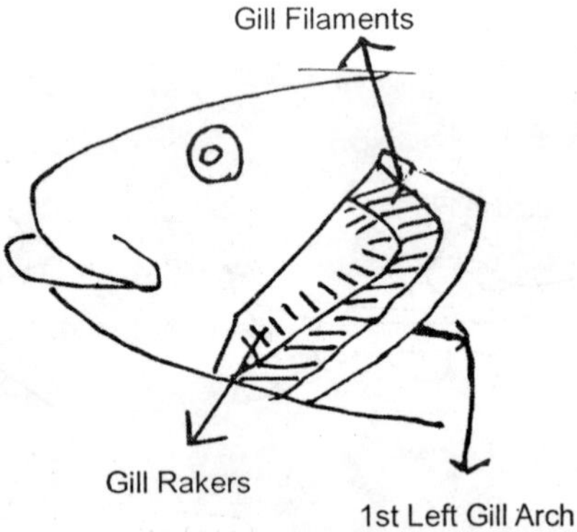

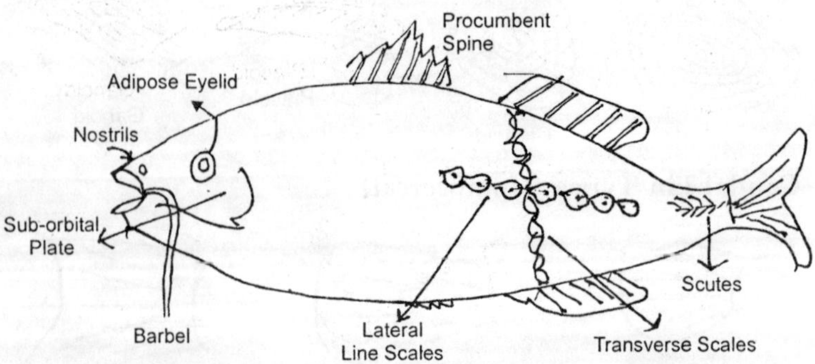

Heterocercal Caudal Fin in Cartilaginous Fishes

SKELETAL SYSTEM IN FISHES

Skeleton: A. Exoskeleton B. Endoskeleton

Funtion:

Invertebrates →	**PROTECTION SUPPORT**	Vertebrates
Exoskeleton →		← Endoskeleton
Scales, bony scale-plates, Fin rays and connective tissues		Notochord, skull, vertebral column and ribs, intramuscular bones, fin supports, girdles.

Table 2

The Skeleton of a Perch

- Axial skeleton
  - Skull
  - Vertebrate
  - Ribs
- Appendicular skeleton
  - Fin supports (dorsal, anal & caudal)
  - Pectoral girdle
  - Pelvic girdle

I. Dermoskeleton of Head

A. **Bones of skull roof:** nasal, frontal, parietal, epiotic, squamosal and post-temporal.

B. **Circum-orbital bones:** lacrimal, sub-orbital, dermo-sphenotic and pterotic

C. **Jaws and palate bones:** premaxillae, maxillae, dentary, angular, vomer, palatine, ectopterygoid, pterygoids and metapterygoid.

D. **Opercular shield:** Preopercular, Opercular, Interopercular and Subopercular.

II. Endoskeleton of the Head

A. **Cranium:** Parasphenoid, basioccipital, exoccipital, supraoccipital, ethmoid (lateral), mesethmoid, epiotic, opisthotic, prootic and sphenotic.

B. **Endochondral jaws and suspensorium:** Hyomandibular, symplectic, quadrate and articular.

C. **Hyoid apparatus and gill arches:** ceratohyal, epihyal, interhyal, hypohyal, urohyal and six branchiostegal rays, gill apparatus has: hypobranchial, ceratobranchial, epibranchial, pharyngobranchial and basibranchial.

Vertebrae: Centrum, neural arch, neural spine, pre-and post-zygapophyses, haemal arch.

Median fins: Spines, pterygiophores in dorsal and anal fins.

In caudal fin: epurals, hypurals, and urostyle.

Pelvic fin and girdle: basipterygia (pelvic plates)

Pectoral fin and girdle: Cleithrum, supracleithrum, postcliethrum, scapula, coracoid and radials.

FISH SCALES

Skin of most fishes is armoured by *scales* which are *dermal* in origin.

Scales are: -protective structures

-help in fish identification and classification
-used for age determination of fishes

-help determination of spawning frequency in some fishes

-also assist in the rough estimation of growth

Types of Scales

1. Placoid scale: (sharks, rays and skates)

- enamel like ectodermal cap (vitrodentine)
- cap underlain by dentine with pulp cavity having dentinal tubules
- has a disc-like dermal basal plate with a cusp projecting outward from it through the epidermis

2. Cosmoid scale: (Crossopterygians and lungfishes)

- thinner, harder, outer enamel layer (vitrodentine)

- layer beneath is hard and non-cellular (cosmine).
- a vascular mid-layer with perforate bony substance (isopedine)
- growth is at the edge of the scale from beneath; no living cells cover the surface

3. **Ganoid scale: (Reedfish and Gar; also in sturgeons and paddlefish)**
   - outer layer is a hard inorganic-salt substance (ganoine)
   - beneath this is a cosmine-like layer
   - innermost lamellar bony layer is isopedine
   - grows at the edges from underneath and also on the surface
4. **Bony-ridge scale: (In most bony fishes; carps (Cycloid), perches (Ctrenoid).**
   - thin and translucent; outer surface with bony ridges that alternates with depressions.
   - inner part or plate with layers of connective tissue
   - growth on both outer surface and from beneath
   - a nuclear central zone present (in most).

Modification of Scales

Jaw teeth (sharks); spines in dorsal fin (spiny dogfish: *Squalus*)

Saw teeth (saw shark; Pristle); stinger on the tail (sting ray: *Dasyatidas*)

Scutes (belly-scutes of herrings); lancets (Surgeon fish: *Acanthurus*)

Dermal bones (skull of bony fishes); soft-rays (Lepidotrichia: *Bony fishes*)

Armature (Sturgeons: *Acipensuridae*)

Perforated lateral line scales (as hardened channel for lateral line.

Sensory canal: Many bony fishes.

FISH MUSCLES

Types of Muscles

A. **Smooth muscles:** Muscles of digestive tract, arteries, reproductive and excretory ducts and eyes.

B. **Cardiac muscles:** Heart muscles.

C. **Skeletal muscles:** Muscles of head, trunk, median fins (protractor, retractor and lateral) and paired fins (abductors and adductors).

From viewpoint of attachment:

A. **Skeletal:** Striated

B. **Non-skeletal:** Smooth and cardiac

From functional point of view:

A. **Voluntary:** Striated skeletal muscles stimulated by motor fibres and is under conscious cerebral control. They are packed in bundles (fascicles) and most are connected to bones by tendons.

B. **Involuntary:**

a. **Smooth:** Non-striated, smooth muscles are slow-acting and is under the control of autonomic nervous system.

b. **Cardiac:** Tireless muscles of vertebrate heart combine characters of both skeletal and smooth muscles. It is fast-acting and striated like skeletal muscles but contraction is under involuntary control (autonomic) like smooth muscles.

Lateral skeletal musculature:

A. **Architectural type:** e.g. lampreys and hagfishes.

B. **Piscine Type:** Living cartilaginous and bony fishes.

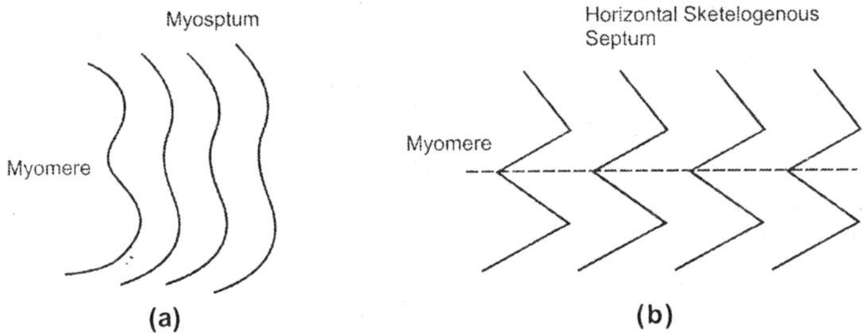

Fig. 17 : (a) Cyclostomine type, (b) Typical Piscine.

5

Locomotion in Fishes

The water is a highly resistant medium to move through. Still some fish moves very fast e.g. 43 km/hr or even exceeding 110 km/hr for brief moments.

This is possible only because the density and non-comprehensibility of water offer great help for forward thrust. Water medium also offers another advantage: since the density of water is only slightly less than of living tissue, fishes are almost perfectly supported and need expend little energy overcoming the force of gravity. Consequently, swimming is actually the most economical form of animal locomotion.

For example: Energy cost per kg body-wt. For travelling 1 km for: Salmon (swimming) = 0.39 Kcal. Gull (flying) = 1.45 kcal. Squirrel (walking) = 5.43 kcal.

Fins (excepting pelvics) are the primary organs of propulsion, the caudal fin is of greatest importance in swimming. But individuals from which all fins have been removed can still swim quite effectively. Yet fins are important for locomotion.

The propulsive mechanism of a fish is its trunk and tail musculature (axial) which is composed of zigzag W-shaped muscle bands (laterally) on the surface but internally deflected forward and backward in a complex fashion.

An eel is a very flexible fish and moves by alternate contraction of the myotomes on either side. The undulatory locomotion has two components:

1. **Thrust:** used to overcome drag and propels the fish forward, and
2. Lateral force: tends to move the fish's head deviate from the course in the same direction as the tail.

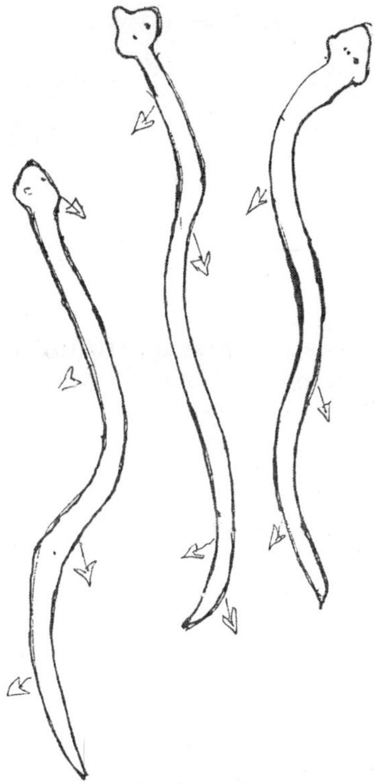

Fig. : Over Simplified Diagrm of Undulating Locomotion

As an ell-like fish moves, a series of alternate flexures back along the body in the sequence a-c. It pushes against water in the areas of flexures indicated by the arrows (from gosline).

6

Fish Blood

Fish blood consists of **plasma (fluid)**, **erythrocytes** and **leucocytes** and is composed of Na, K, Ca, Mg, Cl, P and Fe. Volume of blood in Osteichthyes is 3% body-weight and in Chondrichthyes is 6.6% body-weight.

Plasma contains dissolved minerals and gases, some products of metabolism, enzymes, antibodies and special secretions.

Erythrocyte (RBC): Red colour of blood is due to red blood corpuscles (RBC). Antartic 'ice fishes' lack RBC and have colourless blood also. Blood of young eels (*Leptocephalus* larvee) is colourless.

Fish RBC are oval, nucleated and long diameter ranges from 7 to 36 microns. Hemoglobin (red pigment) present in RBC is the respiratory pigment. It is a protein and iron-porphyrin compound (mol. wt. about 67,000). Bony fishes have 4 heme units (active O_2 - carrying) in the hemoglobin (Hb) molecule. The number of RBC and O_2 - carrying capacity etc. vary with life habits and activity levels of fishes. For example:

Table 3

Species	*O_2-carrying Capacity (% by volume)*	*RBC (per cu.mm)*	*Fe content (mg/100cc)*
Goosefish (Lophius sp.) sedentary habit	5	867,000	13.4
Mackeral (Scomber Sp.) very active	16	3,000,000	37.1

Hemoglobin has high affinity for O_2 which increases with decrease of temperature RBC (Hb) takes up (loading), transports and releases (unloading) oxygen[1]. Also depends on CO_2 content of water. As the partial pressure of CO_2(PCO_2) increases, higher

O_2 - tension of blood (PO_2) is needed for loading, while the O_2 - tension of blood for unloading is lowered and release of O_2 to tissue cells is affected. Also, an increase in H^+ - ion concentration has similar effects.

Leucocyte

White blood corpuecles (WBC) are ovoid or spherical in shape. Their number vary between 20,000 to 150,000 per cu.mm. WBC are of two types:

A. Granular		B. Agranular
a. Acidophil granulocyte b. Neutrophil granulocyte c. Basophil granulocyte	according to their staining reactions	a. lymphocyte (most numerous) b. monocyte (macrophage) c. thrombocyte (blood clotting

Note

1. **Loading tension of blood:** that partial pressure of O_2 (PO_2) at which Hb of a fish is 95% saturated with O_2.

   Unloading tension of blood: that PO_2 at which Hb is 50% saturated with O_2.

7

LYMPHATIC SYSTEM IN FISH

What is Lymph?

Lymph is a fluid the composition of which, in general, rather is similar to that of blood except that RBC are lacking and protein content much lower than blood plasma.

Lymphatic System

It is a closed system composed of delicate tubes (lymph vessels) and large space (lymph sinuses) and conveys fluids away from the tissues and finally returns to the main blood stream.

In fishes, lymph volume is about four times to the blood volume. The lymphatic circulation alone is available in the bulk of myomeres (There are no significant blood vessels in white muscles).

There are several lymph propulsors or "lymph hearts" along the length of major lymphatic vessels which aid lymph return during breathing movements.

The lymph of majority of bony fishes finally flows into the main blood stream through the anterior lymph sinus which opens into the cardinal veins. The true lymph heart appears ventral to the last vertebra on the hypurals. The heart is double chambered and valved.

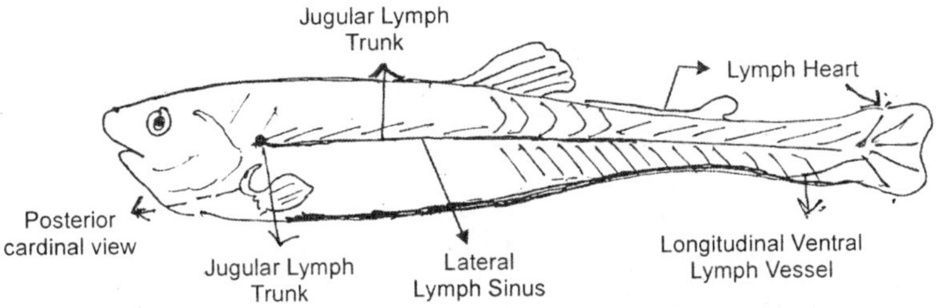

Fig. : Lateral View of Superficial Lymph Vessels and The Lymph Heart of a Young Brown Trout.

8

SWIM BLADDER, NEUTRAL BOUYANCY & ITS MODIFICATIONS

All fishes are slightly heavier than water because their skeletons and other tissues contain heavy elements that are present only in trace amounts in natural waters. To keep from sinking, sharks must always keep moving forward in the water. The asymmetrical (heterocercal) tail of a shark provides the necessary tail lift as it sweeps to an fro in the water, and the broad head and flat pectoral fins act as angled planes to provide head lift. Sharks are also aided in bouyancy by having very large livers containing a special fatty hydrocarbon called *squalene* that has a density of only 0.86. The liver thus acts like a large sack of buoyant oil that helps to compensate for the shark's heavy body.

By far the most efficient flotation device is a gasfilled space. The *swim bladder* serves this purpose in the bony fishes. It arose from the paired lungs of the Devonian freshwater bony fishes when, as we have seen, the alternating wet and dry climate would have made such as accessory respiratory structure essential for life. The primitive lung may have been similar to the lungs found in the existing *Polypterus*, the chondrostean fish of tropical Africa that so resembles the palaeoniscid ancestors of modern bony fishes. Swim bladders are present in most pelagic bony fishes but are absent in tunas, most abyssal fishes, and most bottom dwellers, such as flounders and sculphins.

By adjusting the volume of gas in the swim bladder, a fish can achieve neutral buoyancy and remain suspended indefinitely at any depth with no muscular effort. There are severe technical problems, however. If the fish descends to a greater depth, the swim bladder gas is compressed so that the fish becomes heavier and tends to sink. Gas must be added to the bladder to establish a new equilibrium bouyancy. If the fish swims up, the gas in the bladder expands, making the fish lighter. Unless gas is removed, the fish will rise with ever-increasing speed while the bladder continues to expand, until the fish pops helplessly out of the water. (This is a very real hazard for divers in helmeted diving suits, who must carefully adjust the air pressure to prevent overinflation or underinflation of their suits).

Fishes adjust gas volume in the swim bladder in two days. The less specialized fishes (trout, for example) have a *pneumatic duct* that connects the swim bladder to esophagus; these forms must come to the surface and gulp air to charge the bladder and obviously are restricted to relatively shallow depths. More specialized teleosts have lost the pneumatic duct. In these fishes, the gas must originate in the blood and be secreted into the swim bladder. Gas exchange depends on two highly specialized areas: a gas *gland* that secretes gas into the bladder and a *resorptive area*, or "oval", that can remove gas from the bladder. The gas gland contains a remarkable network of blood capillaries, called the *rete mirabile* ("marvelous net") that acts to transfer gas, especially oxygen, from the blood to the swim bladder.

The amazing effectiveness of this device is exemplified by a fish living at a depth of 2400m (8000 feet). To keep the bladder inflated at that depth, the gas inside (mostly oxygen, but also variable amounts of nitrogen, carbon dioxide, carbon monoxide, and argon) must have a pressure exceeding 240 atmospheres, much greater than the pressure in a fully charged steel gas cylinder. Yet the O_2 pressure in the fish's blood cannot exceed 0.2 atmosphere-equal to the oxygen pressure at the sea surface.

Physiologists who were at first baffled by the secretion mechanism now understand how it operates. In brief, the gas gland secretes lactic acid, which enters the blood, causing a localized high acidity in the rete mirabile that forces hemoglobin to release its load of oxygen. The capillaries in the rete mirabile are arranged so that the released oxygen accumulates in the rete, eventually reaching such a high pressure that the oxygen diffuses into the swim bladder. The final gas pressure attained in the swim bladder depends on the length of the rete capillaries; they are relatively short in fishes living near the surface but extremely long in deep-sea fishes.

SWIM BLADDER IN FISHES

As an organ of sound production and detection

Underwater Accoustics

Velocity of sound: in water: 1500 m/sec. in air: 330 m/sec. wavelength is almost 5 times that in air.

Since water is about 1,000 times denser than air, more energy input is required to initiate propagation of sound in water. But once propagated, sound energy is transmitted faster in water. Transmission is further enhanced by reflection of sound from the water surface.

A. **Sound production:** Swim bladder acts as sound projector when it is vibrated by attached/contiguous muscles.

a. **Indirect:** By deflation, gas bladder help stridulations of the pharyngeal denticles in grunts. It acts as a resonator, may also function literally as a drum (trigger fish).

b. **By gas expulsion:** Sound is produced when air bubbles are eructed through the pneumatic duct (Minnows, Loaches, Eels)

c. **By extrinsic muscles:** Knowking sound (beats) are produced by vibrations of a pair of large muscles contiguous to the lateral walls of the swim bladder e.g. Drum fish

d. **By intrinsic muscles:** Sonic muscles in the bladder itself may produce rapid volume changes resulting in audible signals (toadfish)

B. **Sound detection:** A fish is essentially transparent to waterbourne sound. Its main acoustic discontinuity is swim bladder which is an effective sonic reflector. The *ostariophysi* possess the lowest auditory thresholds and highest upper frequency limits. The Weberian ossicles transmit auditory signals received from swim bladder to the inner ear.

SWIM BLADDER AS A HYDROSTATIC ORGAN

Density of fish flesh = 1.076, while that of fresh water = 1.0005 and that of seawater = 1.026

The swim bladder acts as a hydrostatic organ by replacing part of the fish body which is heavier than water, with gas. The overall density of the fish is thereby brought closely that of the surrounding water and the fish become neutrally buoyant. This makes the fish completely weightless and expend minimum energy in maintaining its position. The gas bladder is able to adjust the amount of gas in the bladder according to changing hydrostatic demands.

The physeclistous fishes adjust the pressure in their gas bladder entirely through secretion or resorption of gases from or to the blood. The physostomous fishes are also capable of gas secretion or absorption by means of the blood supply, but they are unable to fill the gas bladder initially without access to the atmosphere.

When a fish ascends, the bladder expands and gas is removed; when it descends, the bladder is compressed and gas is deposited. The fish is able to vary its gas contents in such a way that the gas volume is almost constant regardless of the hydrostatic pressure.

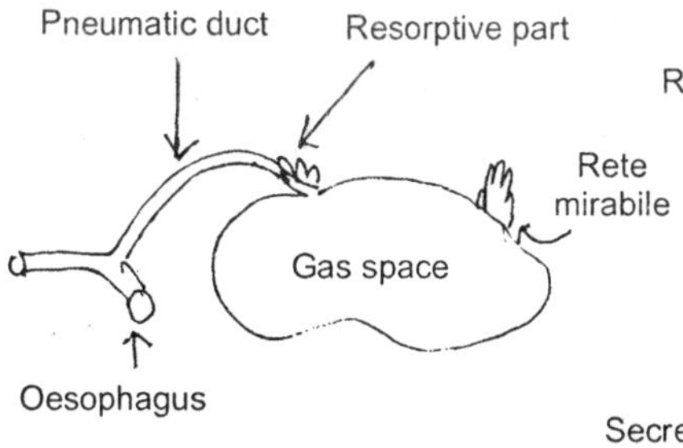

Fig. 13 : Physostomous swim bladder

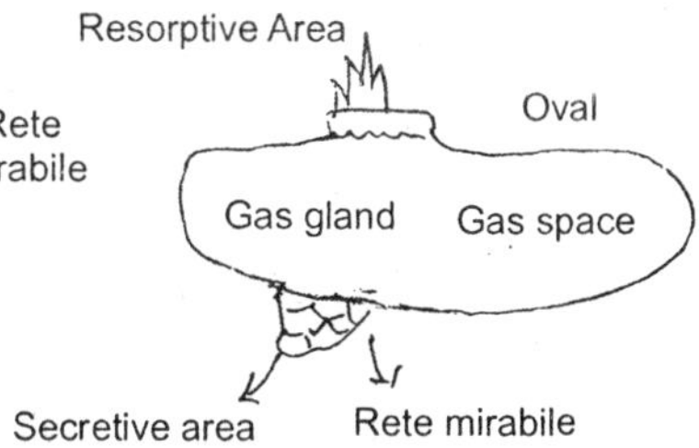

Fig. 14 : Physoclistous gas bladder

How the Gas is Removed?

1. Direct release of gas into the water.
2. Reabsorption of gas into the blood stream.

How the Gas is Deposited?

O_2 present in the gas bladder is derived from molecular O_2 present in the water and is transported by blood hemoglobin. The O_2 reabsorbed is most likely utilized metabolically.

How A Fish Extracts Oxygen From Water

In fishes the folded surfaces of gills far exceed the external body in area. In its respiratory process the fish uses the "*counter-current*" technique well-known to engineers, together with a remarkable "*two pump mechanism*" providing a continuous flow of water through the gills.

Hundred gram fish needs about 5cc of O_2/hr. when at rest and 3 to 4 times when normally active. To remove this O_2 the fish have to pass at least 15 - 30cc of water across its respiratory surface each minute. It is not so easy because the density of water medium is 1000 times greater and 100 times more viscous than air.

Only 1 - 2% of O_2 intake of a man at rest is spent in work done by the respiratory muscles. For fish it is many times greater. In addition, the rate of diffusion of O_2 is 300,000 times slower in water than it is in air.

How then does a fish overcome problems which appear so much greater than those facing terrestrial vertebrates?

Because of the "structure of the respiratory surfaces" and "in the nature of flow across them". The gills are very finely divided series of plates, presenting an enormous surface of contact to the water which passes across them in a single direction, unlike the tidal flow of mammalian lung. An active fish can extract up to 80% of the oxygen contained in the water passing over the gills. Highest figure determined in man is only 25%. Such efficient extraction by fish is due to so-called "*counter current*" relationship between the flow of the blood and of the water, and by the remarkable pumping mechanism which affords a continuous flow of water across the gills throughout the whole respiratory cycle.

How A Fish Extracts Oxygen From Water

In fishes the folded surfaces of gills far exceed the external body in area. In its respiratory process the fish uses the "*counter-current*" technique well-known to engineers, together with a remarkable "*two pump mechanism*" providing a continuous flow of water through the gills.

Hundred gram fish needs about 5cc of O_2/hr. when at rest and 3 to 4 times when normally active. To remove this O_2 the fish have to pass at least 15 - 30cc of water across its respiratory surface each minute. It is not so easy because the density of water medium is 1000 times greater and 100 times more viscous than air.

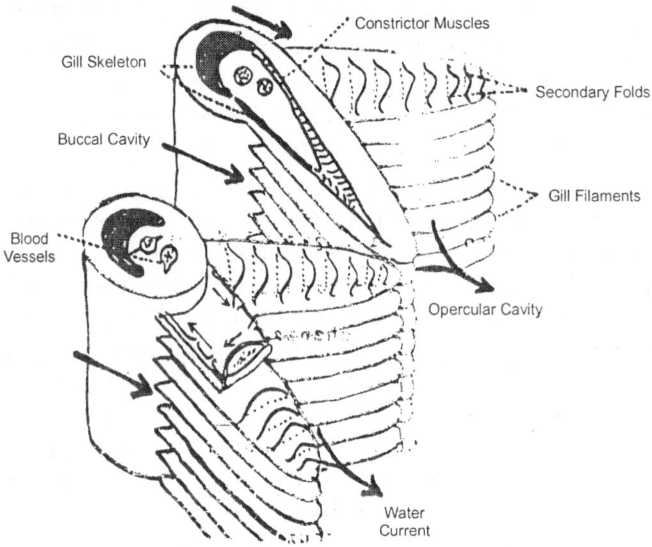

Fig. 15: Two gill Arches(Part) showing filaments of adjacent rows touching at their tips. The blood vessels carrying blood before and after its passage over the Gills are also shown.

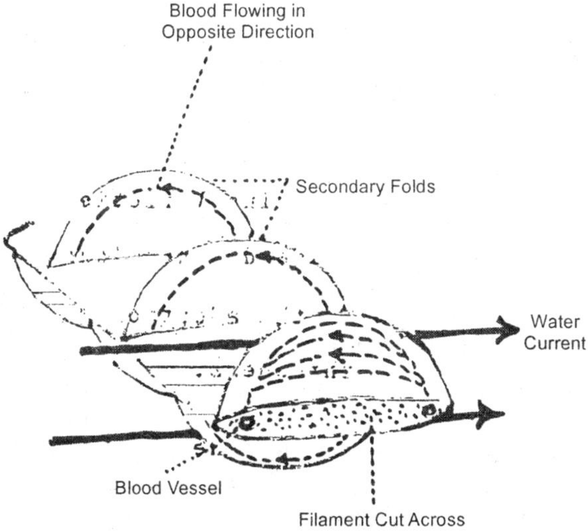

Fig. 16 : Part of a single filament with 3 secondary folds on each side. Flow of blood is in opposite direction to the water.

Only 1 - 2% of O_2 intake of a man at rest is spent in work done by the respiratory muscles. For fish it is many times greater. In addition, the rate of diffusion of O_2 is 300,000 times slower in water than it is in air.

How then does a fish overcome problems which appear so much greater than those facing terrestrial vertebrates?

Because of the "structure of the respiratory surfaces" and "in the nature of flow across them". The gills are very finely divided series of plates, presenting an enormous surface of contact to the water which passes across them in a single direction, unlike the tidal flow of mammalian lung. An active fish can extract up to 80% of the oxygen contained in the water passing over the gills. Highest figure determined in man is only 25%. Such efficient extraction by fish is due to so-called "*counter current*" relationship between the flow of the blood and of the water, and by the remarkable pumping mechanism which affords a continuous flow of water across the gills throughout the whole respiratory cycle.

Counter-current between the Blood and Water

The counter-current principle in many animals where efficient exchange of dissolved materials, gas or of heat is required between two fluids. Van Dam 1938 describes its operation in fish gills. Here it ensures that blood which is about to leave the gills almost fully saturated with O_2 meets water entering with its full O_2 content, while a blood deficient in O_2, entering the gills meets water from which most of the O_2 has already been removed. In this way there is always a greater tension of O_2 in the water than in the adjacent blood, so that O_2 continues to pass into the blood throughout its passage across the gill. When the direction of flow is reversed then only 9% - 51% could be utilized. The blood can carry 10 - 15 times as much O_2 as the same volume of water.

Continuous Flow Across the Gills

When breathing, a fish opens its mouth and water is drawn into the buccal cavity; and passing over the gills it leaves the opercular cavities through slits which appear when the gill-covers expand and come away from the side of the fish. Dr. G. Shelton showed that with the exception of a brief period there is a pressure gradient (that is, the pressure of water in the buccal cavity exceeds that in the operculum cavities) throughout the whole breathing cycle. As a result, water will pass continuously across the gills and so greatly increase the uptake of oxygen.

The mechanism which makes it possible, involves the operation of two pumps slightly out of phase with one another. In the fish the pumping actions are due to changes in volume of the cavities, produced by muscular action.

During the inspiratory phase the buccal cavity expands and water enters through the mouth. At the same time the opercular cavities expand, but water cannot enter through external openings because a thin membrane around the outer rim functions as a valve. During expansion of the opercular cavity the hydrostatic pressure becomes less than that in the buccal cavity and water is drawn across the gills: The opercular cavity acts as a suction pump. During the phase of decrease in volume, which also starts in the buccal cavity, the pressure becomes greater than the external water as the mouth begins to close. It is functionally closed even in those fishes which do not close their mouths completely because of the presence of two thin membranous valves which project just behind the upper and lower lips. During this phase, the increase in pressure within the buccal cavity is greater than that in the *opercular cavities* and water continues to pass across the gills; the buccal cavity acts as, a pressure pump.

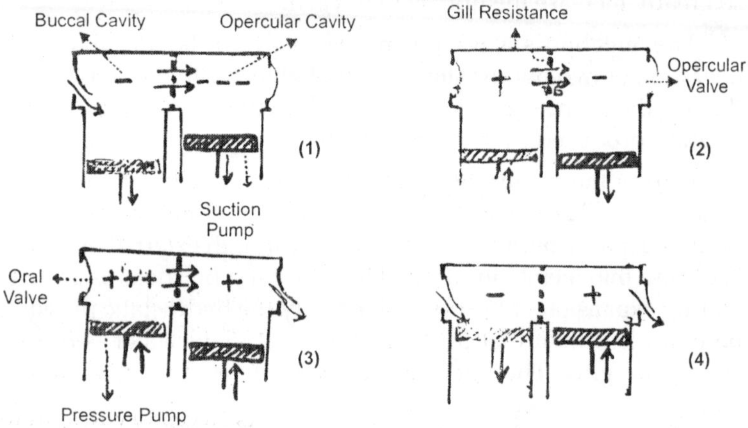

Fig. 17 : Diagrammatic representation of the double-pursping mechanism which maintains an almost continuous flow of water across the gills. The two major phases of the cycle are (1) in which the opercular suction pumps are active and (2) when the buccal pressure pump forces water across the gills. The two transition phases (2) and (4) each takes up only 1/10th of the whole cycle. The pressures in the cavities are given with respect to that of the water outside.

9

ADAPTATIONS FOR AIR-BREATHING IN FISHES

Structures Modified for Air-breathing

1. Slime cover of gills: e.g. spiny eels (*Mastacembelus*)
2. Pharyngeal and opercular wall: mudskippers (*Periophthalmus*)
3. Skin: eels (*Amphipnops*)
4. Modified stomach/intestine: loach (*Misgurnus*)
5. Diverticula of mouth and pharyngeal cavities: *Clarias, Anabas, Ophicephalus* spp.
6. Lining of branchial cavity extended to a pair of elongated sacs: *Saccobranchus*
7. Air-bladder modified:

   a) As lungs in dipnois (Protopterus, Lepidosiren and Neoceratodus)

   b) As accessory respiratory organ: e.g. Gars and Bowfins.

Accessory Respiratory Organs

a) **Ophicephalus striatus:** Two air chambers: ant and posterior. The diverticula carry folded respiratory epithelium with a rich blood supply.

b) **Clarias batrachus:** Opening between 2nd and 3rd gill arches lead to two sacs with highly vascularized arborescent bush-like extensions.

c) **Anaba testudineus:** Consists of a labyrinth (spongy structure) covered with folds or respiratory epithelium and situated in a moist pocket in the gill chamber.

d) **Saccobranchus fossilis:** The gills are fused to form a valve over the orifice of the air sacs which are elongated extensions of the lining of the branchial cavity.

Cyclostomes and Chondrichthyes do not possess gas bladder

Two types of gas bladder in bony fishes:

A. **Physostomous:** G.B. is connected to oesophagus by a pneumatic duct. e.g. soft-rayed bony fishes (trout, salmon, carp)

B. **Physoclistous:** G.B. has no connection with oesophagus. e.g. spiny-rayed bony fishes (perch, seabass)

Functions of Gas Bladder

As Accessory Respiratory Organ:

Many physostomous fishes use gas bladder as a supplementary organ of respiration.

a) Giant redfish (*Arapaima gigas*) of the Amazon River has G.B. modified for aerial respiration.

b) Garfish (Lepisosteus) and the bowfin (Amia) have highly sacculated or alveolar lining in the G.B. These fishes can survive in water devoid of O2 if enabled to swallow air.

10

DIGESTIVE SYSTEM IN FISHES

The ground plan of the digestive tract of fishes follows that of other vertebrates.

The mouth leads to buccal cavity followed by pharynx, esophagus, stomach, pylorus, small intestine and large intestine ending in anus.

The length of the intestine and presence of other features differ with different feeding habits. A long convoluted intestine is characteristic of a herbivorous fish (e.g. carp) since lengthy digestive area is needed for digestion of plants.

Carnivorous fishes generally have shortened intestine.

The sharks and its relative have developed a spiral valve in the intestine in order to lengthen the food passage.

In many teleosts pyloric caeca numbering from 2 or 3 to several hundred are present. Some fishes lack distinct stomach while a few have developed gizzards.

In carnivores, some protein digestion may be initiated in the acidic stomach (due to secretion of HCL and pepsinogen), but in most fishes digestion and absorption generally proceed simultaneously in the intestine.

The function of pyloric caeca is secretion of enzyme (digestive) and absorption of digested foods.

Fish: Major Feeding Types

Predators: Fishes that feed on Macroscopic Animals.

Carnivores: That prey on Myriad of Animal foods from Zooplankton and insect larvae to large Vertebrates e. g. Sharks. .

Piscivores: That mainly feed on other fishes e. g. Seabass, Large Mouth Bass, Snake-Head (Dalag), Grouper (Lapu Lapu).

Larvivores: That feed on Mosquito and other Insect Larvae, e. g. Guppy.

Insectivores: Feeding mainly on Insects, e. g. Some Minnows.

Grazers: Fishes that graze and take by bites often by individual small ones. Food is mainly Plankton or Bottom Organisms, e. g. Blue Gill.

Planktivores: That feed mainly on zooplankters, e. g. Post-Larvae (young fry) of almost all fishes; Also Bighead, Catla.

Strainers: Fishes that feed by straining microscopic organisms mainly Phytoplankton, e. g. Silver Carp.

Suckers: Fishes that suck food into the mouth generally from the bottom mud, e. g. Catostomids (suckers).

Parasites: Fishes that are parasitic on other fishes; They get attached and suck blood and tissue fluid, e. g. Sea Lemprey.

Other Types of Feeding Habits

Herbivores: Fishes that mainly feed on food of plant origin, e.g. Milkfish.

Omnivores: Fishes that feed on both plant and animal food e. g. Tilapia, Common Carp.

Detritivores: That feed mostly on Detritus (Decaying Leaves, Twigs, Organic Debris, Fine Silt and Mud), e.g. Mrigal.

Bottom Feeders: That feed on Bacteria, Fungi, Protozoans, Insect Larvae, Worms etc. living on bottom debris, e. g. Mud Carp.

Macrophytophagus: That feed on higher aquatic plants, grass, e.g. Grass Carp.

Malacophagus: That feed on mollusks (snails etc.), e.g. Black Carp.

Fish

Fishes are cold-blooded aquatic vertebrates which breathe through gills and propel and balance themselves by fins.

They are Taxonomically Divided into:

A. *Cyclostomata*: Jawless fishes e.g. Lampreys and Hagfishes.

B. *Chondrichthyes*: Cartilaginous fishes e.g. Sharks, Rays and Skates.

C. *Osteichthyes*: Bony fishes Carp, Perch etc. Majority of Living Fishes.

Fish as a group has a unique position in lake and stream Ecosystems.

Fish have been always neglected by LIMNOLOGISTS despite their:-

- ECOLOGICAL importance being at the apex of most Aquatic Food Webs;
- COMMERCIAL value, being the main produce harvested from inland waters;
- FISH interact with various levels of the food chains and occupy practically diversified voracious feeding habits;
- FISH inhabit almost all habitats from highest elevation to the deepest ocean bottom and over 1.5 km in inland lake Baikal;
- FISH have acclimatized to temperatures below freezing point at Arctic to over 40° C in some hot springs;
- FISH size vary from the minute (12 mm) Pigmy Goby of the Philippines to the Largest Whale Shark (22.5 m/25 mt) and Giant F. W. Fish of the Amazon River, Arapaima gigas (5 m/200 kg);
- FISH are the most numerous (over 20,000 species) among Vertebrates and have adapted to diverse Ecological and Environmental Conditions.

Food and Feeding Habits of Fishes

Fishes have highly diverse feeding habits involving all different types of feeding; only one broad characteristic prevails, that is *'The Food Is Taken Into The Mouth'*, fish feed on almost all the organisms present in water, such as Zoo- and Phytoplankton, Fil. Algae, Macrophytes, Benthic Organisms; and also feed on fish themselves, Amphibians, Reptiles, Birds and even mammals.

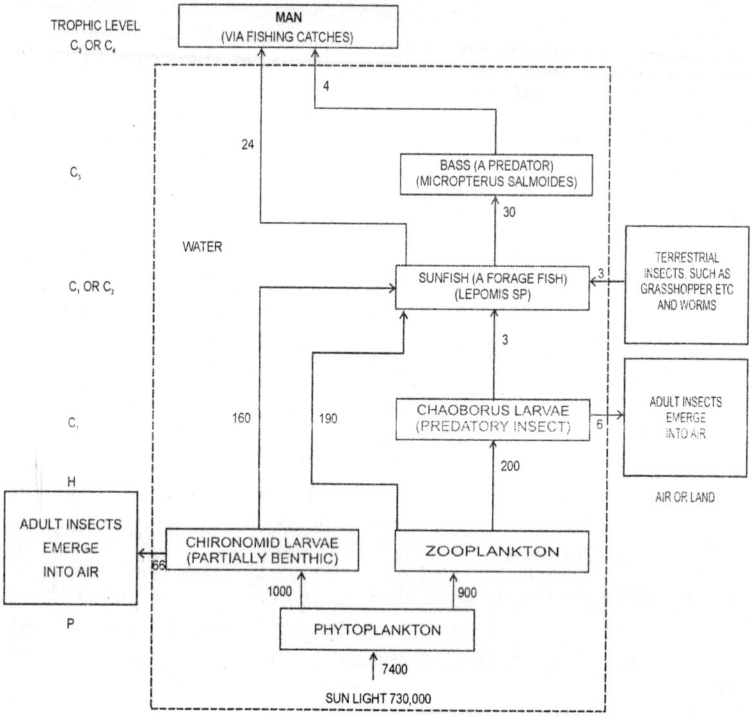

Fig. : Dynamic Food Web for a Small Lake Managed for Fish Production

Values Shown are in Kilocalories per square Metre per year (Uncorrected for Losses in Respiration or Assimilation).

The function of pyloric caeca is secretion of enzyme (digestive) and absorption of digested foods.

11
STRUCTURE AND FUNCTIONS OF FISH KIDNEY

Kidneys (paired) of fishes: Pronephric or mesonephric.

A typical fish kidney has many nephrons each consisting of a 1) renal corpuscle and (2) a kidney tubule.

The renal corpuscle (Malpighian body) is made of a glomerulus- a blood vessel tightly coiled with afferent and efferent arterioles and encapsulated by thin kidney cells (*Bowman's capsule*). The glomeruli vary from large to very small size and even may be lacking in some marine fishes. In freshwater fishes the number of glomeruli may exceed 10,000 in each kidney.

The glomerulus and capsule together act as ultrafilters where blood pressure brings about non-selective dialysis of molecules. Thus, serum proteins are retained in the blood. The excretory fluid undergoes changes on its way through the tubules where glucose, various minerals and other solutes get reabsorbed into blood.

The urine of freshwater fishes contains creatine, creatinine, unidentified nitrogenous compounds (may be amino-acids) and a little ammonia. Urine of marine teleosts also contains trimethylamine (TMO). In addition to nitrogenous wastes, fish kidney also assists in water-salt balance (Homeostasis) by the excretion and retention of certain minerals.

12

Osmotic Regulation in Fishes

OSMOTIC REGULATION IN FRESHWATER AND MARINE BONY FISHES

Freshwater fish Maintains osmotic and ionic balance in its dilute environment (Salt conc. 001 - 005 M) by actively absorbing NaCl across gills (some salts enter with food). To flush out excess water that constantly enters body, glomerular kidney produces a dilute urine by reabsorbing NaCl.

Marine Fish must drink seawater to replace water lost osmotically to its salty environment (1.0 M). Sodium chloride (NaCl) and water absorbed from stomach. Excess NaCl is secreted outward by gills. Divalent sea salts, mostly $MgSO_4$, are eliminated with feces and secreted by tubular kidney.

Diadromous Fishes

When an eel (a catadromous fish) reaches freshwater in its upstream migration it faces problems of salt depletion and overhydration. A young salmon (an anadromous fish) migrating downstream to ocean comes up against these conditions in reverse. So, both anadromous and catadromous fishes must be very efficient in their osmotic adjustment. They have glomerular kidneys which can adjust to differences in urine volumes due to different salinities, and possess gills and oral membranes capable of coping both with the uptake and the secretion of certain ions against existing diffusion gradients.

Euryhaline Fishes (about 10% of all teleosts) are able to pass back and forth with ease between both habitats (freshwater

and marine). They have highly adaptable osmoregulatory mechanisms.

	Characteristics	*Freshwater Fishes*	*Marine Fishes*
1.	Type of osmotic regulator	Hyper-osmotic regulator	Hypo-osmotic regulator
2.	Salt concentration in water	0.001 - 0.005 gram moles per litre (M)	1.0 gram moles per litre (M)
3.	Salt conc. in blood	0.2 - 0.3 M	0.3 - 0.4 M
4.	Water : gain/loss	*Gain water;* excess water goes out as dilute urine through mesonephric kidney.	*Lose water;* compensate loss of water by drinking water
5.	Salt: loss/gain	*Lose salt;* salt absorbing cells of gill epithelium transfer ions of NaCl from water to blood	*Gain salt;* unwanted salts are dispersed by special salt-secretory cells of gill epithelium. Other major salt-ions- Na^+, K^+ and Cl^- are carried by blood to gills for disposal; Mg^{++}, Ca^{++} and $SO4^{--}$ ions left in intestine are either (a) void through feces or (b) a small fraction (10% - 40%) enter blood stream & excreted through kidney.
6.	Formation of urine	By usual filtration resorption method	Kidney excretes $MgSO_4$ and $CaSO_4$ by tubular secretion
7.	Glomerulus	Large functional glomerulus present	Glomeruli lost their impt. and are absent in some marine teleosts.

13
GAMETOGENESIS AND EMBRYOGENESIS IN FISHES

Gametogenesis is the series of transformations that results in the formation of mature gametes (germ cells). Gametogenesis in the testes is called *spermatogenesis* and in the ovaries is called *oogenesis*.

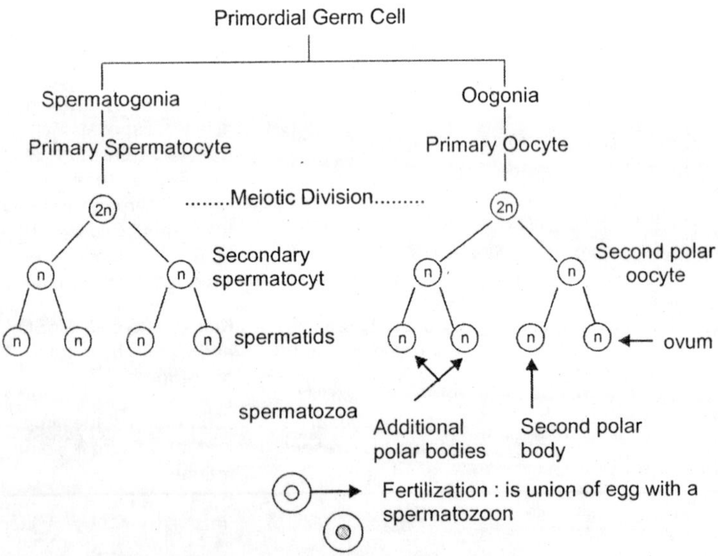

The fertilized egg is called *zygote*.

Embryogenesis: (Development of embryo from the fertilized egg).

Cleavage: The division of zygote into smaller and smaller cells.

A. **Meroblastic:** partial or discoidal cleavage.

B. **Holoblastic:** complete cleavage.

Blastula: a hollow ball of tiny cells (result of cleavages)

Gastrulation: sorting out of cells of blastula into germ layers (ectoderm, mesoderm & endoderm) which form the body organs.

Differentiation: formation of body organs and tissues form the zygote.

Types of Eggs

A. **Isolecithal egg:** small egg with small amount of yolk. Cleavage is typically holoblastic and blastomeres are nearly equal in size e.g. amphioxus.

B. **Telolecithal egg:** eggs with large amout of yolk (50 - 90%).

   1. **Holoblastic cleavage:** later cleavages produce unequal cells e.g. Amphibians.

   2. **Meroblastic cleavage:** cleavages are partial, e.g. fishes, reptiles and birds.

Fish Eggs

Pelagic (floating)/demersal (which sink); spherical/non-spherical; sticky/non-sticky (non-adhesive); oil globules: single/many/none; Yolk: segmented/non-segmented; eggs: with or without filaments; egg membrane: smooth/with structure.

14

Endocrine Organs in Fishes

Interaction of important endocrine glands:

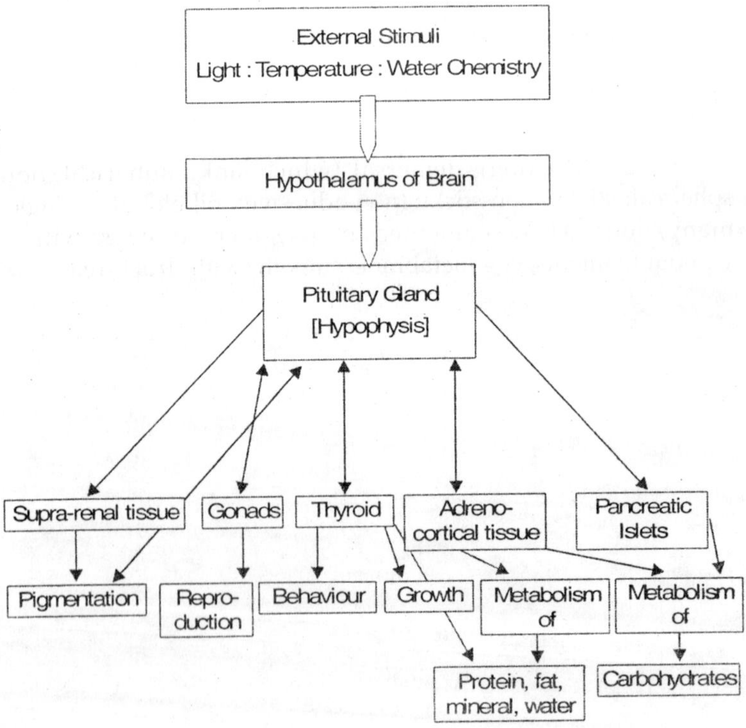

Table : Important Glands, Hormone and their Action

Important Glands	Hormone	Action
1. *Hypophysis* (master gland)		
a. Neurohypophysis	Vasopressin	Osmoregulation by influecing gill permeability
	Oxytocin	
b. Adenohypophysis		
Meso-	Corticotropin	melanogenesis (control cortin secretion)
Meta-adenohypophysis	growth hormone - Somatotropin (STH)	increase in growth
Pro-adeno hypophysis	Thyrotropin	control thyroid secretion
	-Gonadotropin	control secretion of gonadal hormone
	Prolactin Intermedin	Melaongenesis
2. *Thyroid* (Diffused)	Thyroid hormone	Fastens maturity; controls metabolism
3. *Adreno-cortical tissue*	Adreno-cortical	promotes utilzation of stored fats, carbohydrates & protein metabolism
4. *Suprarenals* (Chromaffin)	Epinephrine	Pigment concentration
5. *Pancreatic islets*	Insulin	Carbohydrate metabolism
6. *Gonads* a. Ovaries b. Testes	Estrogen Androgen	Secondary sexual characters; Maturation of gametes.

15

Sense Organs in Fishes

The sense organs in fishes receive physical and chemical stimuli from the environment.

1) Cutaneous Sense Organs (Skin Receptors)

Temperature perception and touch are felt through skin receptors. Taste buds are located in the epidermis of mouth and lips, in the pharynsm snout and barbels.

2) Organ of Vision (the Eyes)

Visual stimuli involve changes in light quality and intensity. The retinal mechanisms have developed towards contrast enhancement and motion perception besides the formation of a clearly focused image.

3) Organ of Smell

Organ of smell is a pouch which opens through nares. The lining of the sac is typically a lamellar sensory epithelium connected by the olfactory nerve with the forebrain.

4) Organs of Hearing

Ampullae — Semicircular Canals

Sacculus — Utriculus

Lagena

a) *Inner Ear*

The inner ear is an organ of hearing (pars inferior) as well as the organ of equilibrium (pars superior). Pars inferior consists of sacculus and lagena (sound receptors). Pars superior consists of semicircular canals, ampullae and utriculus.

b) *The Lateral Line System*

It receives acoustical stimuli. It has widely differentiated sensory structures. It informs the fish of localized disturbances caused by small current by mechanical vibration or by irregular non-violent displacement of the surrounding water.

16

Colouration in Fishes

Colouration in Fish is due to:

Pigments present within chromatophores in the dermal layer of the skin.

The basic pigments are:

-red and orange (erythrophore)

-black (melanophore) and

-yellow (xanthophore).

These pigments are blended to produce other shades (green, brown, etc.). The silvery colour of fish skin is due to guanin which is a waste product of blood, an excretion.

Principal Functions of Colouration

(A) Concealment: The various kinds of concealing colours are:

1. **General colour resemblance** - between fish and the background.
2. **Variable colour resemblance** - ability of fish of change colour.
3. **Obliterative shading** - optical principle is that of countershading.
4. **Disruptive colouration** - concealment by means of camouflage
5. **Coincident disruptive colouration** - a special type of camouflage

(B) Disguise: often relating structure with habit.

Gars (*Lepisosteus*) have the habit of basking motionless near surface, strongly resembling floating twigs.

(C) Advertisement

Some forms of colourating of fishes appear to advertise or to reveal rather than conceal their presence (e.g. darters). Colouration of this type may be of significance for sexual recognition.

Examples:

(A1) Many littoral fishes, (A2) leptocephalus larva or eel, (A3) most finfishes, (A4) scorpionfish, (A5) young angelfish and (B) garfish and needlefish.

17
Bioluminescence in Fishes

Living light (bioluminescence) is present in many mid-water and bottom-dwelling deep sea fishes. This may be due to:

A. Presence of luminous bacteria living on fish in a symbiotic relationship.

B. Self-luminous cells (photophores).

Luminous Bacteria

They inhabit definite but various pockets in the skin often backed by mirroring tissue and equipped with mechanism (shutter type) to regulate the emission of light.

Examples: cods, grenadiers, luminous cardinal fishes, and deep sea angler fishes.

Photophore Cells

The photophores are cells composing glands, the light emission of which is under nervous and/or endocrine control.

Examples: some sharks, electric ray, deep sea scaly dragonfishes and lanternfishes.

Functions of Light Organs

- To illuminate environment to attract food:
- as a means of defense.
- May have value to fishes for recognition, even sexual recognition, especially at night or in the dark depths of the sea.

18

ELECTRIC ORGANS IN FISHES

Fishes are the only members of animal kingdom with body organs that produce electricity. Electric organs are specialized for production of an electric field outside the body. Two groups can be distinguished:

A) Strongly Electric

- Discharge is large
- Fish is painful to handle
- A defensive organ against predators
- An offensive organ for securing food
- Discharge having monophasic pulses; active intermittently.

B) Weakly Electric

- Produces potentials too small for offense and defense
- Functions as part of an electrosensory system
- Detects objects by means of distorsion of the field
- Sensing elements are the electroreceptors.
- Electro-sensory systems can communicate between fishes
- Pulses-Mono-, Di-, triphasic or even more complex; emit continuous pulses of constant size.

The generating cells of electric organs are modified from *muscle fibres* except the *sternarchid family* (Gymnotids). In these the organ is modified from *nerve fibres*. Generating cells are usually flattened (electro-plate/plax).

* *Gymnarchus niloticus* has primitive underwater radar. It sends out electrical pulses; interference with pulses helps them find food or navigate through muddy water where the eyes would be useless.

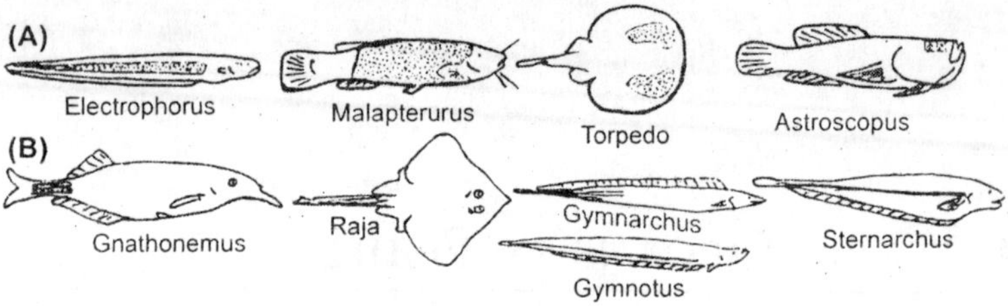

Electric Organs

Fig. : Representative Electric Fish (A) Strongly Electric (B) Weakly Electric.

19
FISH MIGRATION

World Map

Migration is more or less a continuous and direct movement from one location to another. Movement may be:

A) under control of the fish;

B) genetically determined

C) influenced by environmental factors.

Migrations are mostly for:

a. spawning;

b. feeding;

c. fleeing adverse condition.

Factors influencing migratory movements (including homing of fishes) are:

a. **Physical:** bottom materials, depth, pressure, current, tide, turbidity, topography, gradient, temperature and light.

b. **Chemical:** salinity, alkalinity, pH, dissolved gases, odours, taste and pollutants.

c. **Biological:** blood pressure, sexual development, phototaxis, social response, predators, competitors, hunger, food and endocrine state.

Throughout, nervous and endocrine mechanisms are important in orientation and timing.

Spawning Migrations

a) By anadromous Fishes

Pacific salmons (six species of genus *Oncorhynchus*) and the Atlantic salmon (*Salmo salar*) spend their adult lives at sea but return to freshwater streams, where they were born, for spawning. While Atlantic salmon makes upstream spawning runs every year, Pacific salmons make single spawning run, after which they die.

b) By Catadromous Fishes

Freshwater eels (*Anguilla* sp.) make their spawning runs from the coastal rivers of Europe and N. America to Sargasso Sea in N. Atlantic Ocean south-east of Bermuda. Here at depths of 300 m or more, the eels spawn and die. The minute larvae (Leptocephali) begin their journey back to the coastal rivers of Europe and N. America.

The Leptocephali of American eels complete their lifehistory in one year, but European eels take three years to metamorphose to elvers.

Other important migratory diadromous fishes are:

Sturgeons (*Acipensor* sp.; Huso sp.) , American shad (*Alosa sapidissima*), Indian shad (*Hilsa ilisha*), striped bass (*Roccus saxatilis*), Milkfish (*Chanos chanos*), sea lamprey (*Petromyzon marinus*).

Fish Schooling

A school may be defined as ---"A large number of fishes swimming together all oriented in the same direction." Shoaling fishes are usually all of the same length. They appear to disperse into masses of individuals at night and gather again into schools in day time.

Why Do Fish School?

According to Breeder (1954), "A fish school is held together by attractive and repulsive forces." Schooling reduces predation. Studies of fish behaviour suggest that the habit of schooling, whatever its mechanism is, is an advantage to potential prey.

20

ZOOGEOGRAPHY

The study of zoogeography tries to explain why animals are found where they are, their patterns of dispersal and the factors responsible for their dispersal.

A particular species may be absent from a region that supports similar animals for any of the three reasons:

1. There may be barriers that prevent it from getting there.
2. Having gotten there, it may be unable to adapt to the new habitat or compete successfully with resident species and may become extinct.
3. Having once arrived and adapted, it has subsequently evolved into a distinct species.

Three Categories of Distribution:

A. **Worldwide distribution:** reveals broad patterns;

B. **Regional distribution:** dealing with particular groups in selected parts of the world;

C. **Local distribution:** concerned with the geography and ecology of species.

Definition of Key Zoogeographic Terms:

Dispersal: Movements of individuals, particularly to their outward scattering into new habitats.

Autochthonous: Having originated and evolved in the place in question, e.g. N. American darters.

Differentiation: Difference between different species brought about by evolution and other processes.

Dominant: Successful groups relatively numerous with diverse adaptations and wide distribution in many habitats, e.g. the minnows - Fam. Cyprinidae.

Endemic: Confined to a given place or area; occurring nowhere else.

Relict: A survivor; an organism that persists locally after its relatives have died out elsewhere or become extinct, e.g. Coelacanth.

GEOGRAPHIC DISTRIBUTION OF SOME IMPORTANT FRESHWATER FISH GROUPS

***Family Cyprinidae* (*Carps* and *minnows*):** widespread distribution all over the world excepting—S. America, NW Africa, Madagascar and Australian Region.

***Family Catostomidae* (*Suckers*):** Entire N. America: Have only species *Myxocyprinus asiaticus* in China and a recent migrant to NW Kamschkatka (*C. catostoma*) Family Esocidae (Pikes): Family is restricted to North-Temperate Holarctic area only Lungfishes: *Lepidosiren* (1 sp) in S. America; *Protopterus* in Africa and *Neoceratodus* in Australia.

Bowfin (*Amia calva*): only in N. America (1 sp): A phylogenetic relict.

Paddlefish: Only two representatives: One in N. America (*Polyoden*) and one species in China (*Psephurus*).

Family Chaudhuriidae: Only one species (*Chaudhuria caudata*) in Myanmar.

Bichirs (*Polypterids*) are available only in Africa.

21
MULLET

Why Culture Mullets?

Because mullets have the greatest potential of becoming the most important supplier of aquatic animal protein.

Grey Mullets : Most popular cultivated spp. in East as well as in West.

Why? Let us examine their Bio-Data -

- Cosmopoliton Distribution.
- Euryhaline: Can tolerate wide range of Salinity.
- Eurythermal: Can tolerate wide range of temperature.
- High quality and tasty flesh.
- Low position in the food chain (short food chain).
- Respond well to inexpensive methods of fertilization.
- Readily accepts cheap supplemental feeds.

How to Identify Mullets?

- No lateral line.
- 2 short dorsal fins; first with 4 slender spines
- Pectoral fins high on the body
- Eyes often partially covered by fatty tissue (adipose eyelid)

MUGIL: LIZA

Important Genera:

MUGIL	LIZA
• Hind tip of maxilla not curved down elow tip of pre maxilla	Hind tip of maxilla curved down below tip of pre maxilla.
• Adipose cyclid to pupil (in adults)	Adipase cyclid absent or to iris only.

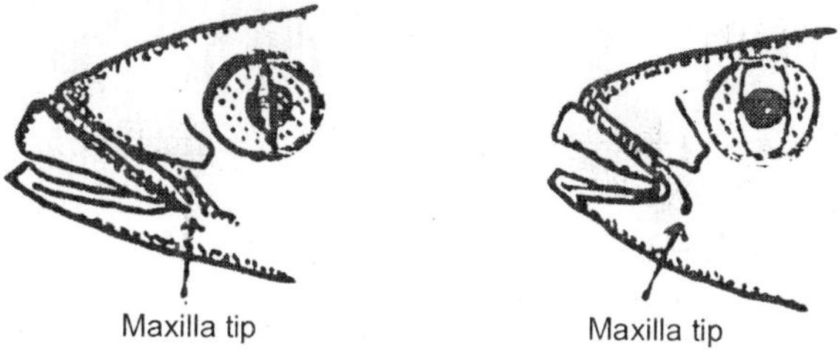

Fig. 36

Some important Commercial Species:

Mugil cephalus
Mugil capito
Mugil chelo
Mugil parsia
Mugil saliens
Mugil dussumicri
Liza troschell:
Liza macroleprie
Liza tade
Liza varigiensis
Liza argents
Mugil curema
Valamugil scheli
Rhinomugil corsula

Grey mullets, especially, *Mugil cephalus* is an ideal species for culture in temperate, tropical and subtropical areas; in sea water, brackishwater and fresh water either in monoculture or in polyculture with other species.

Biology and Ecology of the Grey Mullet, *Mugil Cephalus* L.

1. Introduction

The grey mullet - most popular cultivated species, both in the orient as well as in the West - enjoys wide distribution being euryhaline and eurythermal - high quality flesh - low position in the food chain.

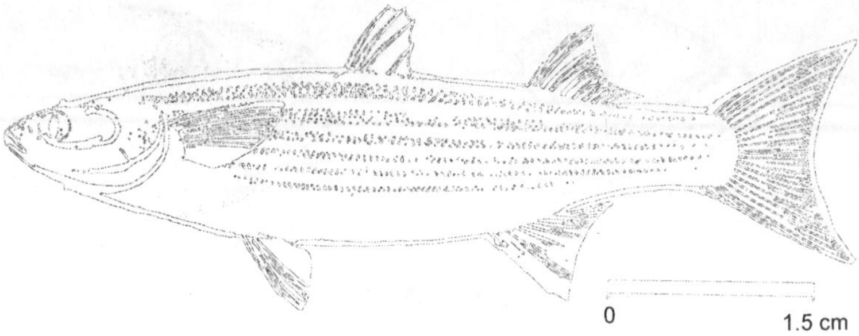

Fig. 24

2. *Taxanomy*

Subclass - Teleostomi.

Order - Perciformes/Mugilformes (Berg)

Suborder - Mugiloidei

Subfamily - Mugilinae

Genus - Mugil

Species - Cephalus.

3. *Morphology and Identifying Characters*

Elongated, compressed body; broad flattended head, mouth small; eyes with adipose eyelid; no lateral line; 2 short dorsal fins - 1st dorsal with 4 slender spines; anal fin with 3 spines and 8 soft rays; hind tip of maxilla not curved below tip of premaxilla. Colour: Olive Green on back, Silvery on sides; 5-7 indistinct longitudinal Brown bars on body.

4. *Distribution*

Cosmopolitan - tropical, sub-tropical and temperate waters.

5. *Habitats*

From open sea to coastal waters, lagoons, mangroves - estuaries - brackishwaters to freshwaters.

6. *Habits*

A fast swimmer, moves in shoals, leaps from the water (hence called jumping mullet) when disturbed.

7. *Migration*

a) Osmoregulatory migration in advanced fry/juveniles from sea to brackish and freshwater - for higher temperature and food.

b) Maturing adults migrate back to sea for final maturation.

c) Spawning migration in the sea.

8. *Temperature and Salinity Tolerance*

Tolerates wide fluctuations of temperature (3 - 35°C) and salinity (0-38 ‰).

9. *Biology of Reproduction*

a) Sexuality - heterosexual, males smaller than females.

b) Sexual maturity - male in 2 years; female in 2-3 years.

c) Gonads - develop in fresh/brackishwater - final maturation in the sea.

d) Fecundity - varies from 0.15 - 2.7 million.

e) Spawning season - varies depending on distribution - spawns more than once in a year.

f) Spawning Ground and Spawning Behaviour - On sandy grounds in bay number of males surround and stay close to a female during spawning.

g) Artificial Propagation - Successful induced breeding by hormone injections - with mullet, carp pituitaries, partially purified salmon gonadotropin - HCG, synahorin with P.G.:Large scale breeding in Taiwan, Hawaii, Israel and also in India.

h) Eggs - Spherical, transparent (0.68 - 0.95mm dia.) with one oil globule (0.29 - 0.38mm).

i) Incubation period: 34 - 44 hours at 22 - 24 C and 30-33‰.

j) Newly Hatched Larva: 2.0 - 3.6mm.

k) Larval Feeding and Growth:

Oyster eggs and larvae, Branchionus with Chlorella (10 day old - 4 mm); copepods and nauplii of Artemia (20 day old - 7 mm); 42 days larval period at 22°C and 32‰; food habit changes at 22- 33 mm.

10. Food and Feeding

Herbivore - feeding on zooplankton and microalgae including epiphytes and benthic forms, decaying vegetation, organic detritus, inorganic sediments and sand particles.

Culture-based Fisheries: Defined As

Production systems in which man conducts commercial fishery production through a series of processes of raising, nursing and catching marine products, using natural or artificial biotic productivity.

- Much higher rate of recruitment of harvestable fish from Spawners.
- Enable expanded exploitation of the basic productivity of the sea.

Japan has established CBF for decades with many commercial species. Such as: Pacific Salmon (Oncorhynchus), Ayu (Pleeoglossus Altivelis), Herring (Clupes), Kuruma Shrimp (Penaeus Japonicus), Scallops (Pecten), Abalone (Halioes) and Red Seabream (Sparus Major).

Activities are: Capture or Culture of Seed, artificial propagation of seed to the juvenile stage, releasing these to the natural environment and capturing them as adults.

Chum Salmon (*O. keta*): 1+42 million fry released in North Japan in 1970. The recovery of adults 2% annual release since 1930.

Ayu (*P. altivelis*): 2,000 - 5,000 million fry released annually (1940-1945). The annual cater during this period was 3,000 - 8,000 tons.

An Example: In recent years, 500-800 million fry of chum salmon are produced every year at a cost of less than 1 yen/fry. The number of returns averaged 2% Fry 0.3-0.5g released. Fish caught 3-4 Kg in 3 to 4 years. 500 million fry costing yen 500 million in 3-4 years would give an yield of 30,000-40,000 tons (value yen 10,000 million) from 10 million chum salmon.

Transplantation and Artificial Recruitment of Anadromous Fish

A total of 2000 million juvenile anadromous fishes are produced every year by artificial means and released in the seas of the World.

The following three genera dominate the artificial recruitment; Oncorhynchus, Acipensor and Salmo. Pacific Salmons, especially *O. keta*, account for more than 90% of the total number of juveniles released for artificial recruitment.

Basin	*Genera*
North Pacific Ocean	Oncorhynchus, Salmo, Morone Saxatilis, Alosor Sapididsslma.
Sea of Okhotek	Oncorhynchus
Sea of Japan	Oncorhynchus
Caspian Sea	Acipenser (Sturgeoms), Huso, Salmo, Oncorhyachus.
Black Sea	Acipensor, Huso puso, Salmo, Morone.
Bultic Sea	Salmo Acipensor.
White sea	Salmo, Oncorhynchus.
Barents Sea	Salmo, Oncorhynchus.
North Sea	Salmo
North Pacific Ocean	Salmo, Oncorhynchus, Morone.
South Pacific Ocean	Oncorhynchus.

I Hear, I Forget.
I See, And I Remember.
I Do, And I Understand.
Ten Thousand Words
Are Not Worth One Seeing.

(Chinese Proverb)

You Give A Man A Fish,
You Feed Him For A Day.
You Teach The Man To Fish,
You Feed Him Every Day.

(Also Chinese Proberb)

Waste Of Time Is The Most Extravagant Of All Expense

Theopharastus

A Man Who Knows How To Make The Most Use Of His Time Makes The Most Of Himself

Maltz

Energy Conversion in the Lake (Fig. 28) is given below:

	Annual Production (Kcal m^{-2})	*% Efficiency of Transfer*
Light	730,000	---
P: Phytoplankton	7,400	1
H: Zooplankton + Chirronomids	1,900	26
C_1: Chacborus Larvae + Sunfish	550	29
C_2: Sunfish (on Chaoborus)	3	
C_3: Bass + Man (on sunfish)	54	11
C_4: Man (on Bass)	4	

The Biomass Pyramid In A Food Chain

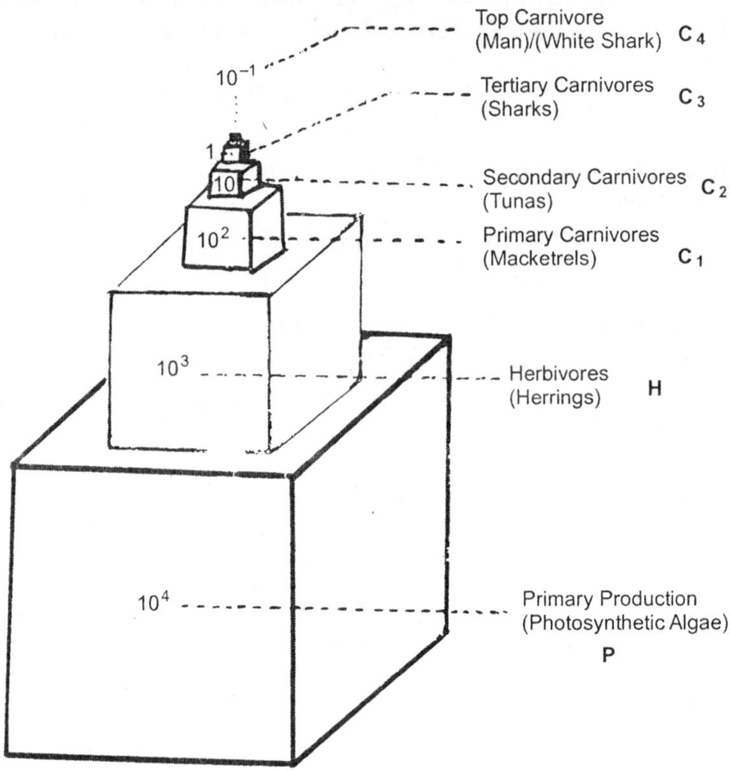

Fig. 29

Energy Available at each Trophic Level is One-tenth (Approx) that Available at Trophic Level next Below

Hence Biomass at each Level is One-tenth (Approx) the Biomass at the Level Next Below.

P.... Photosynthetic Plgae/Phytoplankters are the primary producers.

H.... Herrings are small Herbivorous FISHES.

C_1... Mackerels are larger Carnivorous FISHES.

C_2... Tunas are quite Large Piscivorous FISHES.

C_3... Sharks are very large Carnivorous FISHES.

C_4... There are huge White Sharks which are FISHES and may feed even on Man, the top Carnivore.

Circulation Of Food in a Lake

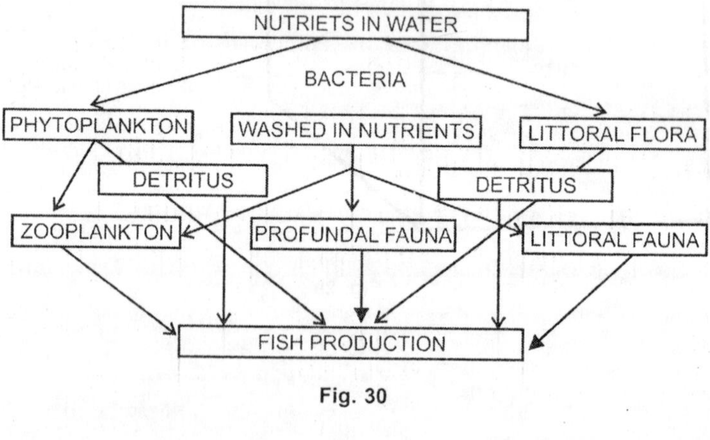

Fig. 30

POSITION OF FISH IN LAKE ECOSYSTEM

Fish is the Final Result of a Complex Biological Cycle in a Lake/Pond/Stream or any Aquatic Ecosystem.

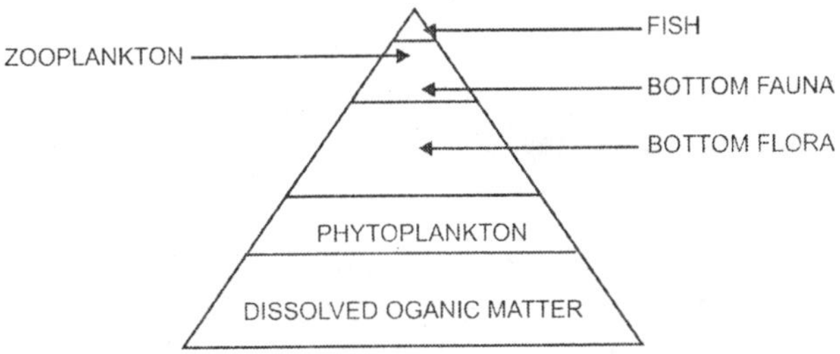

Fig. 31 : Diagram showing weight relationships of Biota and Dissolved Organic matter in Weber Lake, Wisconsin, in Midsummer. The weight of each group is proportional to total area of triangle. (Juday, 1943).

Mass, Energy, or Nutritional Value (kcal M^{-2} $Year^{-1}$)

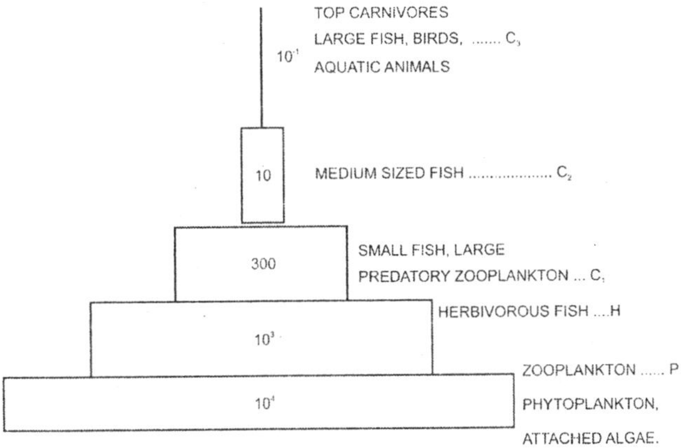

Fig. 32 : Biological Pyramids of Biomass, Energy, and Food value for a typical temperate lake in summer.

Secrets of Achieving High Production Through Running Water Culture

- Abundant Supply of Suitable Water.
- Optimum Rate of Flow.

- Oxygen - Rich Water.
- Effective Elimination of Metabolites.
- High Protein - Rich Complete Diet.
- Quick Growing Cultivated Species.
- Efficient FCR.
- Availability of Seed.
- Optimum Water Temperature.
- Intensive Stocking.
- Feeding and Libitum.

Distictive Features of Fish Culture in the Various Regions of the World due to Difference in Approach to Sociological - Climatic - Biological Factors.

Far Eastern (Asian) System of Fish Culture:

Small-size ponds (0.3 - 2 ha); polyculture of 5-6 species; elaborate system of stocking and careful management resulting in high production with low expenses (use of animal and plant wastes). Here pond is part of an Ecological Complex.

European Fish Culture System:

Big fish farms (50 - 200 ha) producing mainly Common Carp (Monoculture); sometimes with tench and pike. Ponds cropped once a year - growing market size in 3-year rotation mainly on natural food to avoid high expense.

Middle East (Israel) System of Fish Culture:

Moderate size ponds (4 - 6 ha); very intensive management; stocking with two size Carp with either Tilapia or Grey Mullet. Stocking policy: to adjust numbers of each fish group to minimize competitions both intra - and inter - specific, full utilization of resources resulting in high production. (Yashouv, 1968).

How can You Increase Fish Production?

The Secret lies in four 'I's.

1. Increasing the Area of Culture.
2. Improvement of Technology.
3. Intensification of Culture.
4. Introduction of Non-conventional Methods.

Scientific Fish Farming: Proper management

Improperly managed ponds vs. scientific fish farming.

In an improperly managed ponds, the production of fish is a consequence of the natural fertility of the water body and the struggle for existence with the fish community.

As opposed to this in scientific fish farming the fish populations are artificially adjusted and manipulated an their environmental and ecological conditions are suitably altered so that the fish communities will have minimum of stress and strain for existence and maximum favorable conditions to attain their optimum growth.

With these broad guiding principles in view, optimum production of fish could be achieved by any fish farmer.

A MAJOR STRATEGY TO ENHANCE WORLD FISH PRODUCTION

Technological Advances Made in Recent Years Through:

- Polyculture (Multigrade/Mixed Size and Age Stocking)
- Cage and Pen Culture/Culture in Enclosure
- Running Water Culture/Recirculating Tank System
- Integrated Fish/Livestock/Crop Production Systems
- Recycling of Wastes/Sewage-Fed Fish Culture
- Sea-Ranching/Artificial Recruitment/Transplantation
- Controlled Reproduction
- Improving Strains of Fish by Genetic Selection and Hybridization
- Development of High Protein Balanced Feeds
- Introduction of Efficient Exotic Species
- Intensive Research/Technology Generation/Verification and Dissemination
- Development of Technical Manpower
- Development of institutions
- Development of Infrastructures
- Massive Financial Investments
- Suitable Legislations

From

Capture To Culture

Exploring the Limits of Marine Productivity.

Why Culture Fish?

- A good source of cheap protein.
- Easily digestible.
- No fear of Blood Cholesterol.
- Gives higher production than livestock.
- Growing fish is cheaper than poultry/piggery.
- Can utilize all available food in a pond.
- Grows faster.
- Has an efficient FCR (Food conversion ratio).

Food Intake is Spent on

Maintenance **Growth (Most Important For Fish Farmer)**

- For Energy
- For Movement
- For Thermo-Regulation
- For Reproduction

Food used for maintenance is a waste for fish farmers fish spends less energy for maintenance why?

- It does not waste energy in carrying the weight of its body.
- It does not waste energy in heat regulation.

Per Unit of Assimilated Food:

Accumulation of Fish Flesh is:

- 1.5 times more than poultry/pig.
- 2.0 times more than cattle/sheep.

Why? Because of Efficient FCR.

22

Fish Production and Supportive Diagrams

(Evolution of Aquaculture Systems)

Table 6 : Production Of Fish (Arbitrary) In 1 ha. Pond.

Pond	kg/ha/yr	Type of Management
1 ha.	30 kg	Wild Body of Water (No Stocking of Fish: No Management)
1 ha.	100 kg	Man-Made Pond (Unmanaged)
1 ha.	500 kg	Improperly Managed Pond
1 ha.	800 kg	Managed Pond (Inherent Fertility)
1 ha.	2,500 kg	Well-Managed Pond (WMP) (With Fertilization)
1 ha.	5,000 kg	WMP+Fertilization+Supplementary Feeding.
1 ha.	10,000 kg to 20,000 kg	-Do-+Polyculture and Stock Manipulation.
1 ha.	20,000 to 50,000 kg.	-Do-+Aeration, Elimination of Meta Bolites Intensive Feeding.
1 ha.	more than 50,000	+Flowing Water Intensive Stocking Intensive Feeding (Complete Feed).

(A) VERTICALLY INTEGRATED FISH FARMING SYSTEM

A vertically integrated aquaculture industry is one in which all different components of the farming system including hatchery management, grow-out operations, manufacture of feeds, processing as well as marketing of the product are coordinated and controlled by centrally managed systems. It is usually associated with large economic units, large production, large market and large profit.

"It is expected that world aquaculture production will exceed the present volume of the capture fishery before the end of the century and this increase will occur primarily through growth of large-scale vertically integrated aquaculture enterprises" (Weber and Riordan, 1976).

Four major operations can be embraced in a totally integrated fish farming business:

a) Supply of materials and services;

b) Farming itself (i.e. breeding and growing);

c) Processing and

d) Distribution.

Large-scale Aquaculture Industry, particularly in developed countries are after associated with VIFFS.

-Vertical Integration gives scope for centralized planning-

In USA, many factors favour the formation of vertically integrated aquaculture industry. Since this type of organization:-

a) Promotes research and large-scale technology;

b) Induces large capital investment;

c) Utilizes team organization and

d) Assists product development and marketing.

Problems are also many:

1) Species Limitations (market, biological and technological),

2) Sites

3) Feed

4) Manpower

5) Legal and Institutional

6) Financial.

Vertically integrated aquaculture industry is a high risk enterprise and with the high capital investment required, banks and funding institutions are wary of supporting it.

A strategy of Vertical integration can only be justified if it enhances the efficiency of the industry.

VIFFS (VIA)

(1) Support System (2) Production System (3) Marketing System

It Spreads Profit Or Loss Over Different Sections.

- It Embraces -

- Supply;
- Production;
- Processing;
- Distribution and
- SALE.

Vertical integration gives scope for centralized planning. VIFFS is possible mainly in developed countries since the industry is a high risk enterprise needing high capital investment. In USA VIFFS is favoured since such industry:

a. Promotes research and large-scale technology;
b. Induces large capital investment;
c. Utilizes team organization, and
d. Assists production development and marketing.

Constraints:

1. Species Limitations (Market, Biological and Technological),
2. Sites,
3. Feed,
4. Man Power,
5. Legal and Institutional and
6. Financial.

It spreads profit or loss over different sections.

Merits and Demerits of The Two Farming Systems

[A]	[B]
Subsistence Level Fish Farming	***Vertically Integrated Fish Farming***
1. Traditional (Extensive)	Highly Sophisticated (Super - Intensive)
2. Low Inputs	High Inputs
3. Low Technology	High Technology
4. Low Outputs	High Outputs
5. Low Risk	High Risk
Age-Old practice in developing countries and still continuing.	Suitably adapted in some developed countries.

A technologically-developed developing country should take up [A] or [B]?

A via media path is preferable: neither traditional nor too sophisticated. That is -

Small-scale Commercial Level Aquaculture

This type of culture is:

1. Semi-Intensive,
2. Inputs not too high or too low.
3. High Technology
4. Satisfactory Output (Profitable).

In majority of the developing countries, there will be a marked preference for Small-Scale; Labour-Intensive farm of aquaculture for integration with the Rural Development Program. Its success will largely depend on the Support Services that are available. The small farmers would greatly benefit from a `Packages of Technical Assistance, Provisions of Inputs and Institutional Credits' on reasonable terms.

(B) INTEGRATED FISH-LIVESTOCK PRODUCTION SYSTEMS

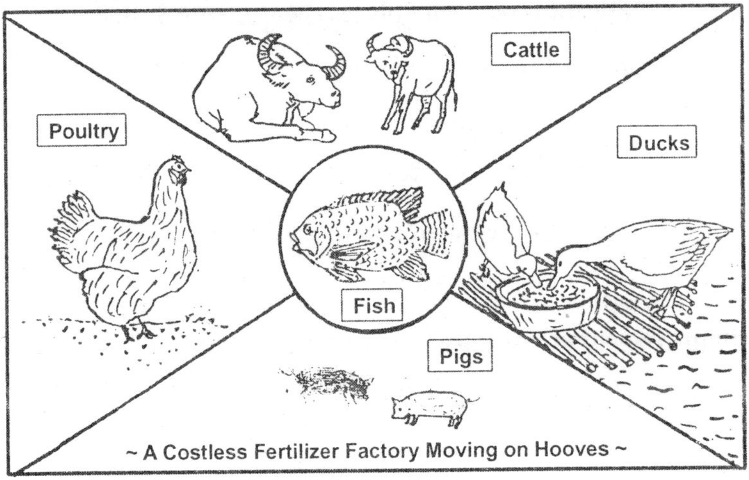

Fig. 32

Manure	Crude Production	N	P_2O_5	K_2O	Organism
Cattle	9.90	0.36	0.16	0.08	17
Pig	19.40	0.5	0.4	0.3	15
Duck	-	1.0	1.4	0.6	-
Poultry	26.90	1.6	1.5	0.9	9

A fat pig produces 2.86 - 2.8 metric tons manure per year. 40- 60 pigs stocked/ha produce manures equivalent to 2 MT of Ammonium Sulfate ($[NH_4]_2SO_4$).

Each duck produces 70 kg fresh manure per year. 500 ducks stocked/ha produce 30 -35 MT manure per year to grow 1500 kg of common carp. Fish yield from manure-loaded ponds without supplemental feeds: 5 - 10 MT/ha/yr.

Vertically Integrated Fish Faming Systems

Three Dimensional Aquaculture

Polyculture of Fish Species : Multigrade System

Monosex Culture : For a -cum Realtory Fish Culture

Cage and Pen Culture

Running Water Fish Culture

Integrated Farming Systems

Introduction of................. Species

Improvement of Stock By Genetic Selection and Hybridization

Table 7 : Polyculture of Asiaticd Carps

Popular Name	*Scientific Name*	*Important Food Items*	*Feeding Zone*
Silver Carp	*Hypophthalmionthys Molitrix*	Phytoplankton	Surface Layer
Bighead	*Aristichthys nobilis*	Zooplankton	Surface Layer
Catla	*Catla catla*	Zooplankton	Surface Layer
Rohu	*Labeo rohita*	Periphyton/ Decaying Veg	Middle Layer (Column Feeder)
Grass Carp	*Ctenopharyngodon idella*	Aquatic Vegetation	Surface/Middle
Tawes	*Puntius gonionotus*	Aquatic Vegetation	Surface/Middle
Mud Carp	*Cirrhinus molitorella*	Bottom Fauna/Detritus	Bottom Layer
Mrigal	*Cirrhinus mrigala*	Bottom Fauna/Detritus	Bottom Layer
Black Carp	*Mylopharyngodon piceus*	Bottom Organisms/ Molluscs	Bottom Layer
Common Carp	*Cyprinus carpio*	Bottom Organisms/ Omnivorous	Bottom Layer
Fishes Used In Polyculture Other Than Carps			
Java Tilapia	*Oreochromis mossamibicus*	Omnivorous	All Layers
Nile Tilapia	*Oreochromis niloticus*	Omnivorous	-do-
Sea Bass	*Lates calcarifer*	Piscivorous	-do-
Chital	*Notopterus chitala*	Small Fish/Insects/ Offal Feeder	-do-
Striped Murrel (Dalag)		Ophicephalus striatus	-do-

∴ Chinese Carp ∴ Indian Carp ∴ Indonesian Carp

Polyculture or Mixed - Fish Farming

Composite Fish Cultures

Comprising

A Number Of Compatible Species Having Different

Feeding Habits They Can Utilize All Available

Food In A Pond to Give Increased Production of Fish

INTEGRATED FARMING

Integrated Fish: Crop: Live Stock Farming Systems.

Fig. 33

In integrated fish/crop/livestock production systems, fish enjoys most of the benefits since the wastes and by-products of crop and livestock are utilized for increasing fish production thereby saving expenditure on fertilizers and feeds.

1. Fish Culture
2. Raising Cattle
3. Poultry Farming
4. Rearing Pigs
5. Duck Rearing
6. Growing Fodder
7. Banana/Fruit Plantation
8. Growing Ipil-ipil/Mulberry
9. Dumping Animal Wastes.

Integrated Agriculture Aquaculture Production Systems

- Reduce Operating Costs - Less Inputs.
- Maximise Farmer's Income.
- Generate useful Fish Protein From wastes.
- Sover waste Management Problem.
- Control Pollution Problem Associated with Livestock.
- Fish Ponds Provide Drinking water to Livestock.
- Help Better Utilization of Skilled Labourers.
- Generate Employment for Farmer's Family/Community.
- Improve Protein Supply/Nutrition to Farmer's Family.
- Provide Self-Sufficiency in Food Supply to Farmer's Community.
- Make Efficient use of Resources and Maximise Yields of Wide Diversity of Food Products.

Socio-economic Issues of Integrated Farming System (Furtado, 1980)

- Information Dissemination
- Detailed Socio-Economic Analysis
- Impact of Technology Transfer
- Marketing and Distribution
- Evaluation of Different Technologies
- Extension Approaches
- Government Policies and Programs
- Validation of Different Systems
- Economic Feasibility of Rice-Fish Culture
- Constraints to Increasing Productivity.

Application of Advanced Technologies to Enhance Food - Fish Production

INDUCED BREEDING

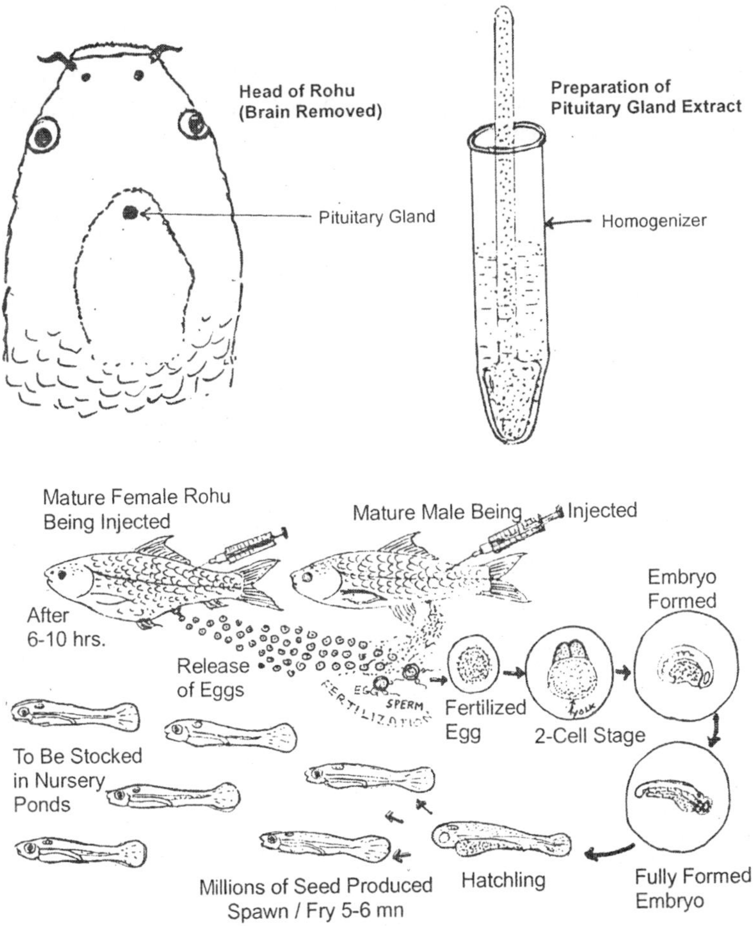

Fig. 34

What is Subsistence Aquaculture?

Dictionary Meaning : Farming or a system of farming that produces a minimum and often inadequate return to the farm operator: economically marginal farming.

OR

Farming or a system of farming designed to provide all or essentially all the goods required by the farm family usually without any significant surplus for sale.

Character:

A. Low Inputs

B. Low Level of Technology (Traditional Methods)

C. Low Output

Yet others argue about the meaning of `subsistence' (It is interpreted differently by different people).

"Subsistence would not be when a person is able to produce some fish for his own consumption. That would not be enough or adequate to afford subsistence for him. Actually he should be able to get an income that would ensure a standard of living comparable that of others in the society where we lives".

Subsistence Level Aquaculture as Sometimes Termed as - Homestead Aquaculture

Subsistence level and small commercial-scale aquaculture currently dominate aquaculture industry in many developing countries of the world.

(F) SEED PRODUCTION : BY INDUCED BREEDING

1. By Manipulation of Environmental Factors
2. By Use of Hormones.

A. Fish Pituitary Extract (FPE/CPE/CPH)

B. Threshold Dose Of FPE/CPE + Synahorin (HCG+Mammalian Pituitary Hormone)

C. Partially Purified Salmon Gonadotropin (SG - G100)

D. Human Chorionic Gonadotropin (HCG) Or HCG + FPE/CPE.

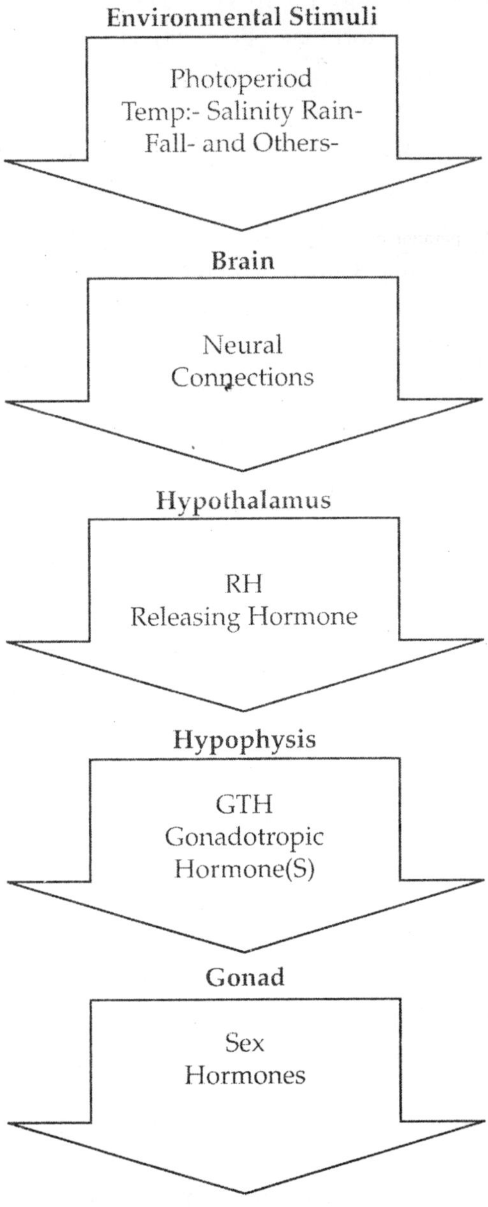

Fig. 35 : Maturation & Release of Sex Cells

Use Of Gonadotropin Releasing Hormone (Gnrh)

E. Luteinizing Hormone - Releasing Hormone (LH - RH)

F. LH - RH Analogue (Decapeptide/Nonapeptide) (LHRH - A)

Release of Matured Gametes : Endocrinological Links

Neuro - Hormones: Regulate GTH Secretion.

A) GNRH (Gonadotropin Releasing Hormone)

B) Dopamine (Gonadotropin Release-inhibiting Factor)

Dopamine Antagonists: Pimozide, Domperidone, Eglonel, Metoclopramide Etc.

$_{m}$GnRH-a (Mammalian GNRH Analogue)

$_{s}$GnRH-a (Salmon GNRH Analogue)

G. LINPE Method: SGNRH-A+Domperidone

Improved Methods of Administration of Hormones:

a) Intrapetitoneal Implantation of Pellet (Hormone)

b) Use of Osmotic Pumps (IP)

c) Pulsatile Release of Hormone

(G) USE OF MISCELLANEOUS HORMONES AND NON-HORMONAL EXTRA BIOLOGICAL SUBSTANCES

Sterodial Hormones, Thyroid Hormone, Progestogens, Prostaglandins, Pheromones Antiestrogens: Clomid, Sexovid, Tamoxifen etc.

Genetic Engineering

By Cloning, Gene Transfer of Growth Hormone and Insulinlike Growth Factors.

Selective Breeding and Hybridization

Interspecific and intergeneric hybrids of Indian Carps have been produced, some showing improved character and faster growth than parent species. Selective Hybridization produces better strains using hybrid vigour.

Many hybrids are sterile and in some cases hybridization leads to production of single sex individuals, such as, all Male progeny produced by crossing Tilapias *O. niloticus* × *T. hornorum*.

Further by genetical research and intensive farming technique, a Red hybrid strain produced by crossing *O. niloticus* × *O. Mossambicus* (Termed as Aquatic Chicken) which is capable of giving production as high as 600-800 mt/ha/yr.

MOLLUSCS

How To Distinguish Molluscs Groups?

Mussel Clam Cockle Scallop Oyster Abalone

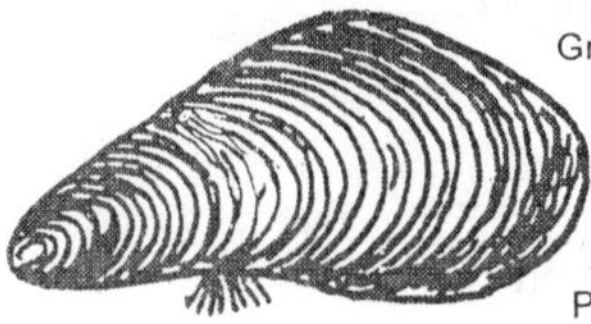

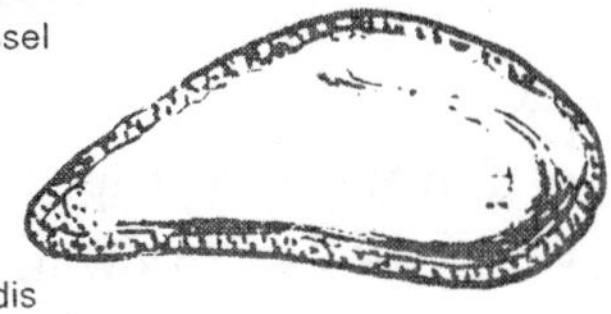

Green Mussel

Perna viridis

MUSSEL: Anterior end pointed - Post. End broader - attaches by byssus threads.

CLAM:

- Clams have long siphons
- Shells of equal size
- Burrowing habit but
- Moves with powerful Foot.

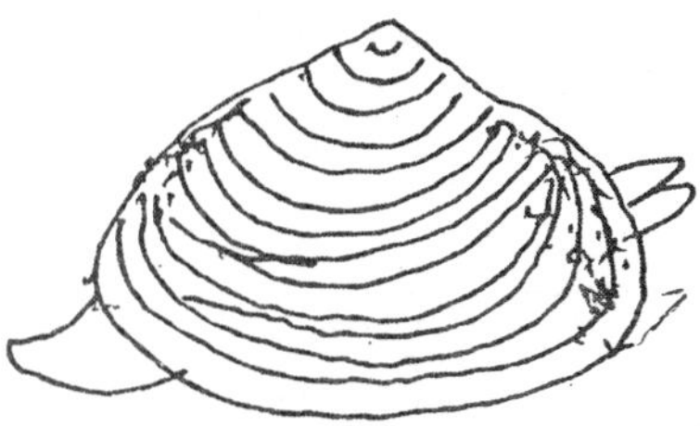

Meretrix Meretrix

COCKLE:

- Heart - Shaped
- Burrowing
- Sedentary
- Can move esily.

Anadora Subcrenata

SCALLOP:

- Fan or Comb-Shaped
- Moves Fast Throwing water on two sides.

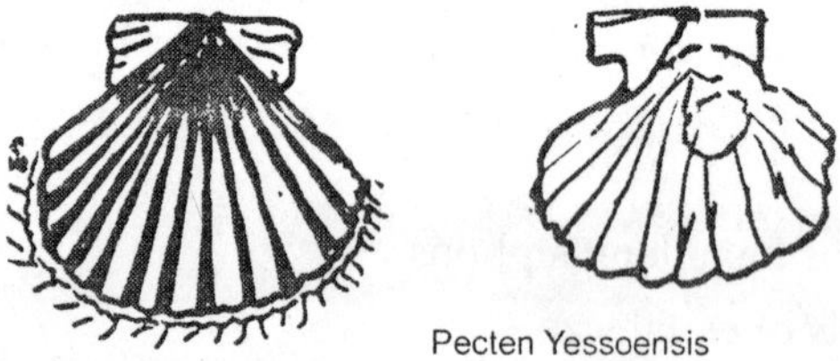

Pecten Yessoensis

OYSTER:

- Sedentary, Unequal Sheels
- Left Shell (Lower) Larger than the Right (Upper)
- Single Adductor Muscle

  Foot small.

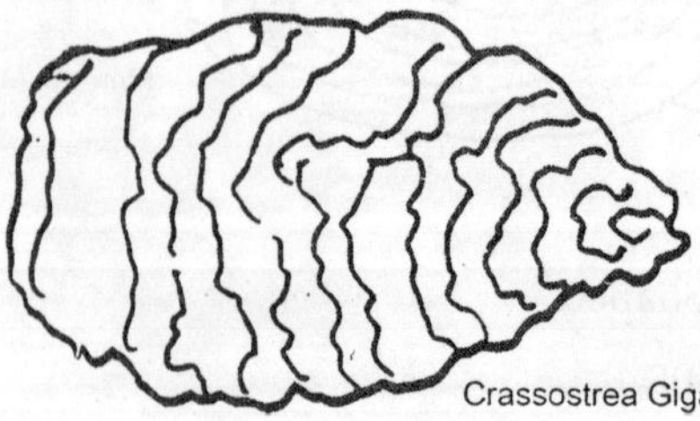

Crassostrea Gigas

ABALONE:

- A Gastropod having single shell
- Holes along the left sides of the shell

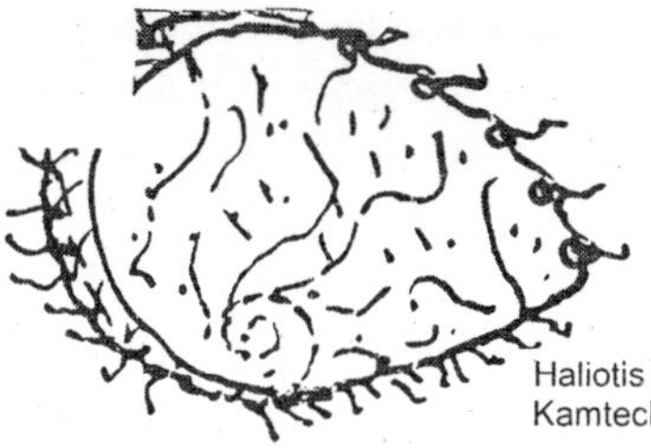

(I) MOLLUSCS OF COMMERCIAL IMPORTANCE

Clam: Cockle: Scallop: Oyster: Abalone

CLAMS: Clams are bivalves with long siphons, equal shells closed by two adductor muscles. A powerful burrowing foot-both marine FW forms - 20,000 edible species - 50 commercial - most inhabit shallow waters-size ranges from 0.1mm to 1.2m usually remains buried from just beneath the surface to depths about 0.6m 1. FW Clams: Glochidium Larva (Parasitic) - Marine Clams: Free Swimming Larva:

2. Veliger - Example: Giant Clam *Tridacna Gigas* (Shells Weigh 263Kg Length 137cm) *Spisula Solidssima* (18cm) *Mercenaria Mercenaria*.

COCKLES: Cockles are Marine Bivalves of the family Cardiidae - 250 species - cockle has 2 Heart - Shaped Shell of equal size and shape with serrated edges of many Radial Ribs - has a strong muscular foot - can move easily - size 1.2cm to 15cm - most species live just below low - Tide Line, some in depth upto 1.6 Km - found in sandy/muddy areas buried to a depth upto 3cm. Example: Giant Pacific Cockle *Trachycardium Quadragenarium Great Heart Cockle Dinocardium Robustum*.

SCALLOPS: (Fan Shell or Comb Shell) Marine Bivales belonging to family *Pectinidae* - 50 *Genera* 400 species - Habitat: *Intertidal Zone* to considerable depths - size 5 -15cm. *Single Adductor Muscle* - can swim rapidly (unlike other Bivalves) by forcibly ejecting water - edges of mantle have man well-developed cell or eyes - Larvas Free Swimming Example: *Giant Or Deepsea Scallop Plecopecten Maggellanicus Atlantic Bascallop Aquipecten Irredians.*

MUSSELS: Mostly comprise freshwater Bivalves of the families union: DAE (750 SPP) *Margaritanidae And Multilidae* - Marine bivalves mostly belong to family *Mytilidae*-distribution nearly all temperate. Tropical seas of the world clinging to rocks in Tidal Zones attached by Byssus thread-valves are elongated, pointed at one end, round at the other e.g. *Mytilus Edulis*(Brown Mussel) *Perna Virdis* (Green Mussel).

OYSTERS: Bivalves of family *Ostreidae* (70 spp. Out of 400 spp. Commercially important) inhabits temperate Tropic Seas-Two Shells Unequal; Left hollow is bigger; Right is thin Float Less Developed -Sedentary type- Foot is small, entirely lacking in some-single large adductor muscle. eg. *Ostrea Edulis, Crassostrea Gigas.*

ABALONES: Large Marine Snails belonging to the family *Haliotidae* 15 species, only about 20 commercial species - there are holes along the left side for respiration - single coiled shell. Example: *Naliotis Discus.*

Molluscs: Phylum Mollusca

(Mollis = Soft)

- Soft Unsegmented Bodies: Usually Bilaterally Symmetrical
- Body surrounded by Thin Fleshy Mantle
- Commonly Sheltered in an External Limy Shell Secreted by Mantle
- Many with Prominent Ventral Muscular Foot
- Mouth with Radula Unique Rasping Organ (Except Bivalves)

Class : Amphineura : Foot Flat, Shell with series of Flat Plates

1. Aplacophora: Wormlike Solenogastre (Neomenia)
2. Monoplacophora: e.g. Neopilina
3. Polyplacophora: e. g. Chiton

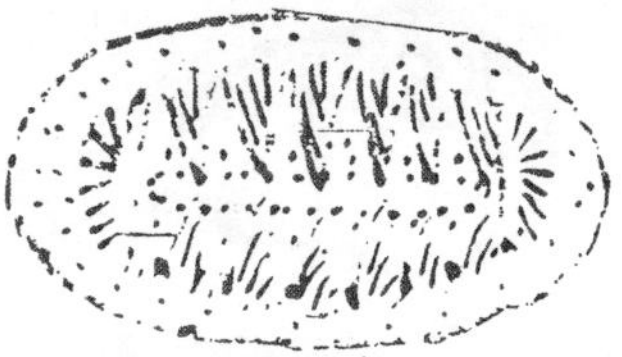

Chiton

Class : Scaphopoda

- A small group with conical shell
- Open at both ends: Dentalium (Tooth Shell or Tusk Shell)

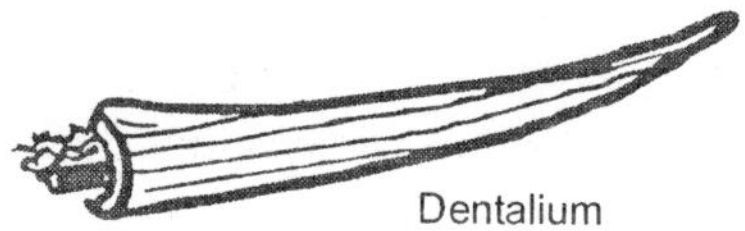

Dentalium

Class : Pelecypoda (Bivalvia)

(The largest commercially important Molluscan Group)

- Foot Spade or Axe Shaped
- Two Lateral Shells Hinged Dorsally

Examples: Clam, Cockle, Scallop, Mussel, Oyester etc.

Clam

Class : GastropodaCLASS GASTROPODA

(Bilateral symmetry is frequently obscured by a secondary change of torsional nature)

- Broad Ventral Flat Foot
- Shell Single and Usually Coiled

e.g.: Snail, Slug, Abalone etc.

Land Snail

Class : Cephalopoda

- Body divided into a large head and sac like trunk
- Foot surrounds mouth forming tentacles/siphons
- Fastest moving and most developed Molluscs.

e. g.: *Octopus, Squid, Nautilus* etc.

Sepia

SOME NOTABLE FACTS ON ICHTYOLOGY (IN DIAGRAM)

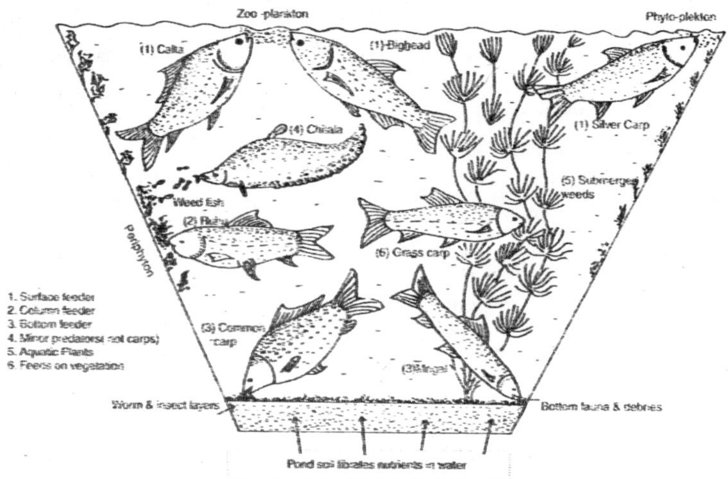

Fig. 1

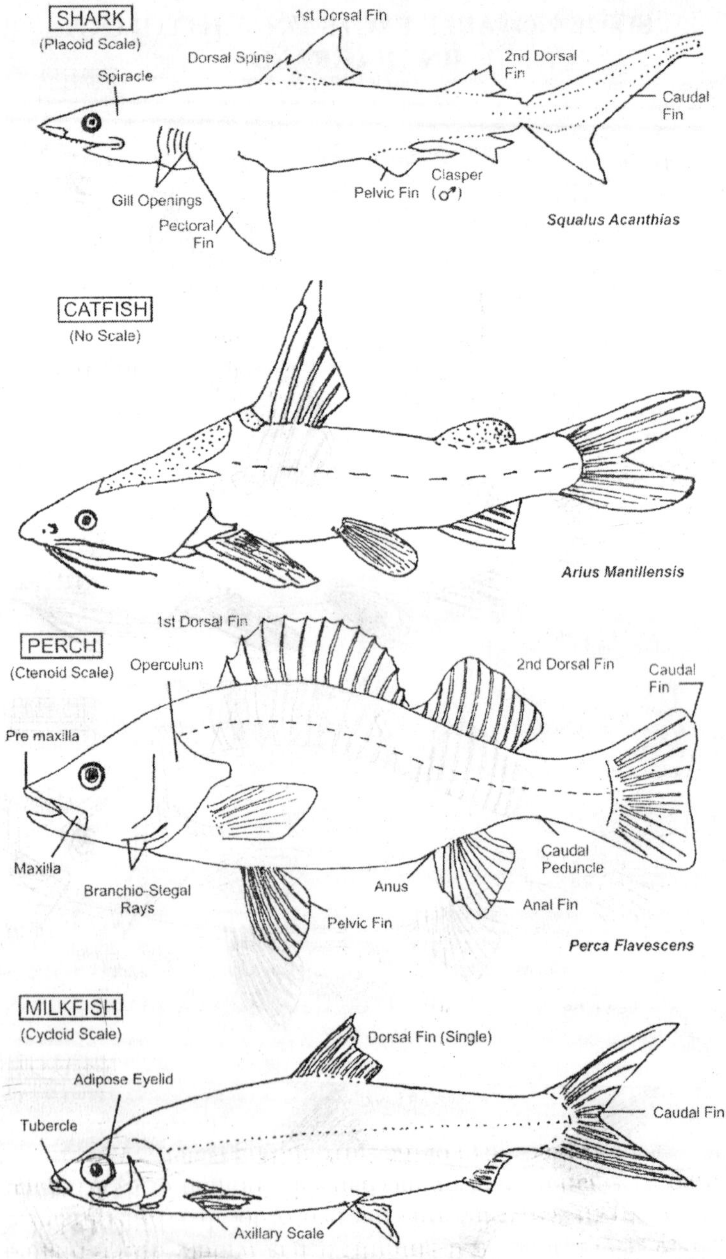
SHARK
(Placoid Scale)
1st Dorsal Fin
Dorsal Spine
Spiracle
2nd Dorsal Fin
Caudal Fin
Clasper (♂)
Pelvic Fin
Gill Openings
Pectoral Fin
Squalus Acanthias
CATFISH
(No Scale)
Arius Manillensis
PERCH
(Ctenoid Scale)
1st Dorsal Fin
Operculum
2nd Dorsal Fin
Caudal Fin
Pre maxilla
Maxilla
Branchio-Stegal Rays
Caudal Peduncle
Anus
Anal Fin
Pelvic Fin
Perca Flavescens
MILKFISH
(Cyclcid Scale)
Dorsal Fin (Single)
Adipose Eyelid
Tubercle
Caudal Fin
Axillary Scale

Fig. 2

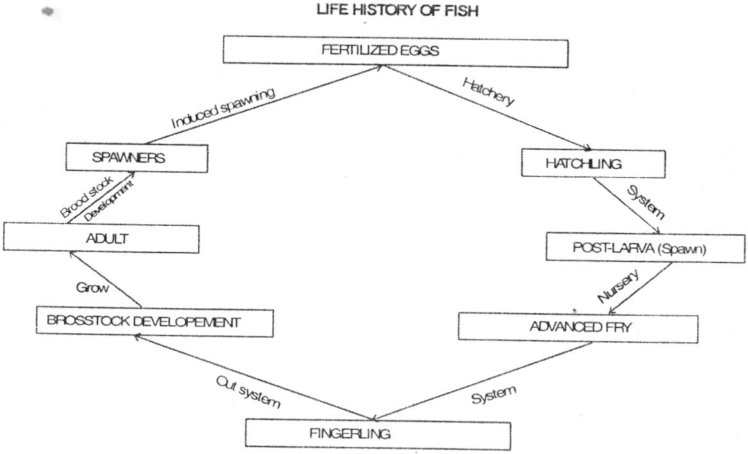
LIFE HISTORY OF FISH
FERTILIZED EGGS
Induced spawning
Hatchery
SPAWNERS
HATCHLING
Brood stock Development
System
ADULT
POST-LARVA (Spawn)
Grow
Nursery
BROSSTOCK DEVELOPEMENT
ADVANCED FRY
Cut system
System
FINGERLING

Fig. 3

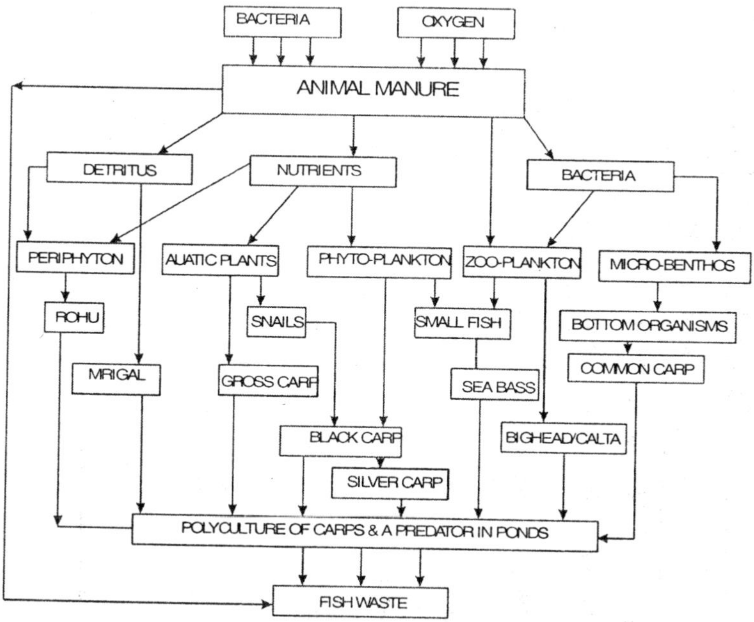
BACTERIA
OXYGEN
ANIMAL MANURE
DETRITUS
NUTRIENTS
BACTERIA
PERIPHYTON
AUATIC PLANTS
PHYTO-PLANKTON
ZOO-PLANKTON
MICRO-BENTHOS
ROHU
SNAILS
SMALL FISH
BOTTOM ORGANISMS
MRIGAL
GROSS CARP
SEA BASS
COMMON CARP
BLACK CARP
BIGHEAD/CALTA
SILVER CARP
POLYCULTURE OF CARPS & A PREDATOR IN PONDS
FISH WASTE

Fig. 4

Culture of *P. Monodon* (Tiger Shrimp) : Grow-Out System

Intensity of Culture	*Estensive*		*Semi Intensive*	*Intensive*		*Super Intensive*
Items	*Traditional*	*Improved TRI*		*Ordinary*	*Sophisticated*	
Size of Pond/Tank	5 to 100 ha	3-20 ha	1-3 ha	0.1-1.5 ha	0.05-0.5 ha	0.01-0.05 ha
Depth of Water (m)	0.3-0.6 m	0.3-0.6 m (1m In trench)	0.8-1.0 m (1-2 m in trench)	1.5 m	1.5-2.0 m	2.0 m
Fertilization (Pond)	None of Organic	Organic & Inorganic	Organic & Inorganic	Mostly Synthetic	Synthetic	Usually not practiced
Supplementary Feeding	No	No	Tash fish & formulated feeds	Intensive Feeding with High Protein Diet		Complete Balanced Feed
Water Management	Tidal	Hardly any Mainly Tidal	Water Exchange 5-15% daily	W.E. Upto 30% per day	W.E. 50-80% per day	W.E. 90-100% daily
Stocking Rate	3,000-5,000 fry/ha less than 1/m^2	1-2/m^2	2-5/m^2 Mxm. 5-15/m^2	20-50/m^2	100-250/m^2	500-600/m^2
Production	50-200 Kg/ha/yr	200-500 Kg/ha/yr	500-1000 Kg/ha/yr Mxm. 1-5 tons	5-10 Tons/ha/yr	10-40 Tons/ha/yr	Over 40 Tons/ha/yr
Remarks	Very Low	Improved	Gives Best Return	Can be Tried	Risky	Very Risky

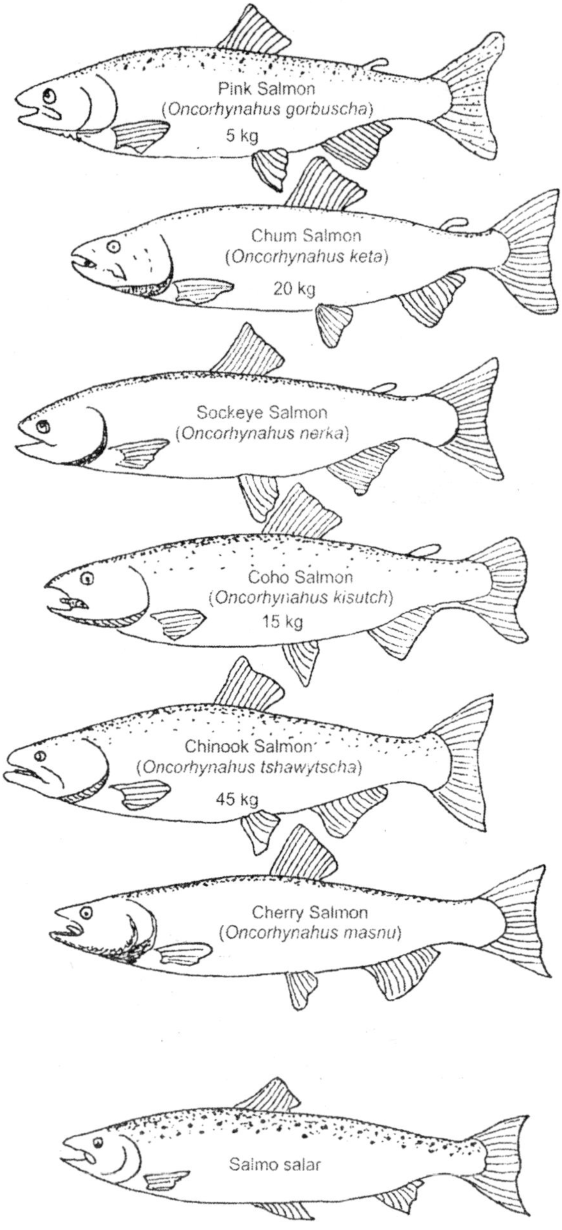
Pink Salmon
(Oncorhynahus gorbuscha)
5 kg
Chum Salmon
(Oncorhynahus keta)
20 kg
Sockeye Salmon
(Oncorhynahus nerka)
Coho Salmon
(Oncorhynahus kisutch)
15 kg
Chinook Salmon
(Oncorhynahus tshawytscha)
45 kg
Cherry Salmon
(Oncorhynahus masnu)
Salmo salar

Fig. 5

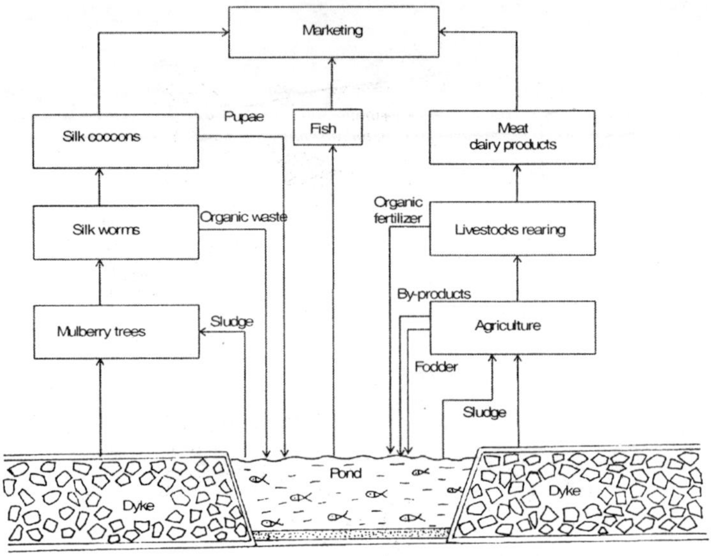

Fig. 6 : Integration of fish culture

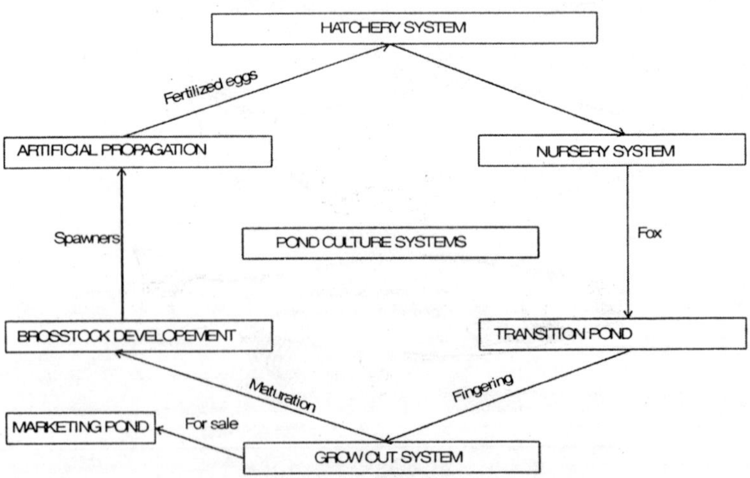

Fig. 7 Polyculture

Reference Books on Ichthyology

1. *Ichthyology:* K.F. Lagler, J.E. Bardach, R.R. Miller. J. Wiley and Sons.
2. *An Introduction to Ichthyology:* P.B. Moyle & J.J. Cech. Prentice-Hall.
3. *Readings in Ichthyology* : M.S. Love and G.M. Cailliet. Goodyear Pub.
4. *Special Ichthyology* : G.V. Nikol'ski.
5. *Systematic Ichthyology* : Greenfield, David Wayne
6. *Biology of Fishes:* C.E.Bond: Saunders Co, Philadelphia
7. *The Biology of Fishes:* H.M.Kyle: Biotech Books, Delhi, India.
8. *An Introduction to the Study of Fishes:* Albert Gunther, Delhi.
9. *The Secret Lives of Fishes:* B.L. Gordon.
10. *The World of Fish:* Franco de Carli: Abbeville Press N.Y.
11. *Fishes of the World:* A. Cooper: Bantam Books, N.Y.
12. *The Life of Fishes:* N.B. Marshall: Universe Books, N.Y.
13. *How Fishes Live:* P. Whitehead: Elsevier, London
14. *Living Fishes of the World:* S.E. Herald: Garden City, N.Y.
15. *FAO Species Identification Sheets for Fishery Purposes (Eastern Indian Ocean and Western Central Pacific),* FAO Rome (Vols. 1-IV)
16. *Checklist of Philippine Fishes:* A.W. Herre; Research Report 20, USA.
17. *Key to the Families of Common Commercial Fishes in the Philippines:* A.F. Umali: Research Report 21 Fish & Wildlife Ser., USA
18. *Fish Processing Handbook for the Philippines:* A.C.Avery: Res. Rpt. 26.
19. *English & Local names of Philippine Fishes:* A.W. Herre:

19. *English & Local names of Philippine Fishes:* A.W. Herre: Circular 14
20. *Commercial Fishes of the Philippines:* Nobort Rau & A. Rau
21. *A Guide to Philippine Flora & Fauna,* Vol. 1: P.V. Conlu
22. *Fishes of Laguna de Bay:* A.N. Villejo Jr.
23. *Freshwater Fishery Biology:* K.F. Lagler: WMC Brown Co.Dubuque, Iowa
24. *Fishes and their Ways:* C.J. Hylander: McMilan Pub. Co. N.Y.
25. *Functional Morphology of Teleostean Fishes:* W.A. Gosline, Honolulu
26. *How to know the Freshwater Fishes:* S. Eddy & J.C. Underhill
27. *Classification of Fishes, both recent & Fossils:* L.S. Berg, Michigan
28. *The Genera of Fishes:* D.S. Jordan
29. *Principles of Animal Taxonomy:* G.G. Simpson
30. *Principles of Taxonomy:* L.J. Gier
31. *Classification of the Animal Kingdom:* R.E. Blackwelder
32. *Aquarium Fishes of the World:* J.H. Prescott: Bantaw Book Inc., N.Y.
33. *Exotic Tropical Fishes:* Axelrod *et al.*
34. *The History of Fishes:* Norman
35. *The Life Story of the Fish:* His Morals and Manners: Brian Curtis
36. *Fishes of the World:* Hans Hvass: Wilhelm Eigner, London
37. *An Age of Fishes: Development of the most successful vertebrate:* J.O.I. Spoczynska: Scribner, N.Y.

BIBLIOGRAPHY

(Scientific papers, Technical Reports, Lectures, Extension Pamphlets and Technical books published/Mimeographed)

H. Chaudhuri 1951, Notes on the embryonic development of the 'Transparent Goby' *Gobiopterus chuno* (Hamilton), Proc. Nat. Inst. Sci., India. 17(4) : 247-252.

H. Chaudhuri 1956, Effect of pituitary injections and other factors on the reproduction of pond fishes (MS-Dissertation. 1955). ABSTRACT Published in *Ala. Poly Inst. Bulletin.*, 1956.

H. Chaudhuri 1959, Notes on external characters distinguishing sex of breeders of the common Indian Carps. *Sci. & Cult.*, 25 (10) : 258-259.

H. Chaudhuri 1959, Experiments on hybridization of Indian carps. Paper read at 46th. Session of Indian Science Congress. *Proc. Indian Sci. Congr.*, 46 (4) : 20-21.

H. Chaudhuri 1960, Experiments on induced spawning of Indian carps with pituitary injections. *Indian J. Fish.*, 7(1) : 20-49.

H. Chaudhuri 1961, Contributions to the techniques of pond fish culture in India (D. Phil. - Dissertation).

H. Chaudhuri 1961, Spawning and hybridization of Indian Carps. Proc. Pacific Sci. Congr., 10th Pacific Science Congress. (Hawaii. 1961 (Abstract only).

H. Chaudhuri 1962, Induced breeding and development of common catfish, *Ompok bimaculatus* (Block)., *Proc. Indian Sci. Congr. 49*(3).

H. Chaudhuri 1962, Breeding and life-history of *Rasbora elanga* (Ham.). *Proc. Indian Sci. Conr., 49*(3).

H. Chaudhuri 1962, Breeding of *Puntius sarana* (Ham.) and observations on its life-history and bionomics. *Proc. Indian Sci. Congr., 49*(3).

H. Chaudhuri 1963, Induced spawning in Indian carps (Chandrakala Hora Memorial Lecture). *Proc. Nat. Inst. Sci. India* (B), 29(4) : 478-487.

H. Chaudhuri 1964, Introduction of exotic species of fish and their effect on the culture of indigenous species of India. Seminar Inland Fish Dev., U.P., Lucknow (Printed by Bureau of Extension, Education & Training, Lucknow) : 83-92.

H. Chaudhuri 1964, Techniques of induced breeding in fishes with special reference to major carps of India. Seminar Inland Fish. Dev., U.P., Lucknow (Printed by Bureau of Extension. Education & Training, Lucknow) : 65-73.

H. Chaudhuri 1967, FAO Report to the Government of Burma on the development of Fish Culture. Based on the work of H. Chaudhuri. FAO.UNDP/TA Inland Fishery Biologist. Rep. FAO/UNDP (TA), (TA 2298) : 14 p.

H. Chaudhuri 1968, Breeding and selection of cultivated warmwater fishes in Asia and the Far East - a review. FAO Fish. Rep. (44) : 4 : 30-66 (Paper presented at FAO World Symposium on warmwater Pond Fish Culture. Rome, Italy, 18-25 May, 1966).

H. Chaudhuri 1969, Breeding habits of cultivated fishes. Paper presented at FAO/UNDP Regional Seminar on induced breeding of cultivated fishes. Calcutta-Cuttack-Bombay, July 15-Aug. 18, 1969 FRI/IBCF/4 : 13 P.

H. Chaudhuri 1969, Methods adopted for inducing breeding of fishes. Paper presented at FAO/UNDP Regional Seminar on Indued Breeding of cultivated fishes. Calcutta-Cuttack-Bombay, July 15-Aug. 18, 1969. FRI/IBCF/8:16p.

H. Chaudhuri 1969, Techniques of hypophysation in Indian Carps. Paper presented at FAO/UNDP Regional Seminar on Induced Breeding of Cultivated fishes. Calcutta-Cuttack-Bombay, July 15-Aug. 18, 1969. FRI/IBCF/11 : 15 p.

H. Chaudhuri 1969, Induced spawning of cultivated fishes. Indian Farming, Fisheries Number, Dec. 1969. 19(9) : 71-74.

H. Chaudhuri 1971, FAO Report to the Government of Burma on Fish Culture Development. Based on the Work of H. Chaudhuri. Inland Fishery Biologist (Fish Culture). Rep. FAO/NDP (TA), (2954) : 45 p.

H. Chaudhuri 1971, Stocking pond management in composite fish culture. Paper presented at FIRST ALL - India Coordinated Res. Proj. Workshop on "Composite Fish Culture" CIFRS (ICAR) Cuttack 1971 : 14 p. (mimeo).

H. Chaudhuri 1971, Culture of air-breathing fishes. Paper presented at First All-India coordinated Res. Proj. Workshop on "Air-breathing fishes for culture in swamps." CIFRS (ICAR) Cuttack, 1971 : 12 p. (mimeo).

H. Chaudhuri 1971, Fish hybridization in Asia with special reference to India. Lecture delivered at Seminar/study tour in the USSR on "Genetic Selection and Hybridization of cultivated fishes. 19 April - 29 May 1968. FAO FISH. UNDP (TA) Rep. FIR/UNDP (TA) 2926 : 151-159.

H. Chaudhuri 1972, Production of fish seed by induced breeding. Seminar on "Production of quality fish seed for fish culture." CIFRI (ICAR) Barrackpore, West Bengal. Nov. 1 & 2, 1968. Proc. Published on Sept. 1972 : 270-282.

H. Chaudhuri 1973, Fertility of hybrids of Indian major carps and preliminary studies on the F_2 generation of carp hybrids. *J. Inland Fish. Soc. India*, 5 (Dec.) 1973 : 195-200.

H. Chaudhuri 1974, Selective breeding and hybridization of carps. Lecture delivered to Summer Institute on "Inland Aquaculture." CIFRI (ICAR), Barrackpore, India, 1974 : 16 p. (mimeo).

H. Chaudhuri 1974, Report of the Government of Fiji on the 'Grass carp project.' CIFRS (ICAR) Cuttack, 1974 : 25 p. (mimeo).

H. Chaudhuri 1975, Experiments on the effects of organochlor insecticides Aldrin, Dieldrin and Endrin on fish and other pond organisms. J. Inland Fish. Soc. India, 7 (Dec.) 1975 : 189-203.

H. Chaudhuri 1975, Fish seed production - A prerequisite to intensive freshwater fish culture. Indian Farming : Inland & Coastal Aquaculture Number, 25 (6) Sept. 1975 : 13-15.

H. Chaudhuri 1975, Activity Report of Hiralal Chaudhuri, Consultant to IDRC, (Canada) Carps (Malaysia) Project, freshwater Fish. Res. Stn. (MADRI) Batu Berendum, Malacca. 1975 : 14 p. (mimeo).

H. Chaudhuri 1976, Use of hormones in induced spawning of carps. Paper prsented at 13th Pacific Science Congress. Vancouver B.C. 18-30 Aug. 1975. J. Fish. Res. Board Canada, 33 (4) pt. 2, 1976 : 940-947.

H. Chaudhuri 1978, Report on the consultancy visit of Hiralal Chaudhuri to IDRC Carps (Malaysia) Project. Freshwater Fish.

Res. Stn. (MARDI), Batu Berendum, Malacca, Malaysia, 1978 : 40 p. (mimeo).

H. Chaudhuri 1978, Report on the consultancy visit of Hiralal Chaudhuri to IDRC Aquaculture (Sudan) Project, Fish. Hydrobio, Res. Stn., E1 Shagarra, Khartoum, Sudan, 1978 : 35 p. (mimeo).

H. Chaudhuri 1982, Pond Culture Techniques (pp. 515-548). Contribution to the Agriculture Yearbook "Research in animal Production." Published by ICAR, New Delhi, India, 1982 : 697 p.

H. Chaudhuri 1983, Status and problems of carp hatchery and nursery management. Published in "A compilation of SEAFDEC AQD Technical papers on "Milkfish and other Finfishes." 1:81-90.

H. Chaudhuri 1984, Fisheries and Aquaculture in Lao PDR. Paper presented at the Auburn Symposium on Fisheries and Aquaculture, Auburn U. Auburn, AL. Sept. 19-23, 1984 : 1 p. (mimeo).

H. Chaudhuri 1984, Rehabilitation of fish seed farms and fish culture development in Lao PDR - an interim final report to FAO/UNDP, 1984 : 25 p. (mimeo).

H. Chaudhuri 1985, Rice-Fish Culture. Paper presented at the International Rice Research Conference, 1-5 June, IRRI, Los Bafios, 1985 : 35 p. (mimeo).

H. Chaudhuri 1985, Rice and Fish Culture, Asian Aquaculture, 7(10) : 4-7.

H. Chaudhuri 1985, Integrated fish - crop - livestock farming systems. Lecture delivered to the trainees attending "Orientation Cum-training Program on improving income and employment potential of rice farming systems at IRRI, 1985 : 18 p. (mimeo).

H. Chaudhuri 1986, Consultancy Report on "Applied research investigation and technology development in Fish Culture for Lao PDR. 1986 : 21 p. (mimeo).

H. Chaudhuri 1987, Final Report of the programme and progress of the NACA Sixth Training Course for Senior Aquaculturists in Asia and the Pacific Region 1987 : 16 p. with 17 Appendices : 56 p. (mimeo).

H. Chaudhuri 1989, Report on the consultancy visit to Ellah Lakes Ltd. Fish Farm Project, Obrikom, Rivers State, Nigeria, 1989 : 15 p. (mimeo).

H. Chaudhri & K.H Alikunhi 1957, Observations on the spawning in Indian carps by hormone injections, *Curr. Sci 26* (12) : 381-382.

H. Chaudhri & S. M. Banerjea 1965, Report on the fisheries of Manipur, with special reference to the development of Takmu Beel area of Loktak lake. Misc. contribution No. CIFRI, Barrackpore, West Bengal, 1965 : 31p.

H. Chaudhri & U. Hla Tin 1971, Preliminary observations on the survival and growth of Chinese carps in Burma. IPFC Occasional paper 71/5, Regional Office for Asia & the Far East, BKK, 1971: 16p.

H. Chaudhri & R.D. Charkrabarty 1974, Intensive fish farming and development of technique of composite culture of Indian and exotic fishes. Lecture delivered at Summer Institute on Inland Aquaculture. CIFRS (ICAR) Cuttack, 1974: 10 p. (mimeo).

H. Chaudhri & K. Janakiram 1975, Control of aquatic weed by moth larvae. *Nature*. 253 (5486) Jan. 3, 1975 : 40-41.

H. Chaudhri & R.D. Chakrabarty 1975, Method demonstration to educate fish Culturists in management techniques for developement of fish culture in India. paper presented at 13th Pacific Science Congress, Vancouver B. C., Canada, Aug. 18-30, 1975: 12p. (mimeo).

H. Chaudhri & S.D. Tripathi, 1976, Problems of warmwater fish seed production. Presented at FAO Technical Conference on Aquaculture, Kyoto, Japan, 1976. Published in "Advances in Aquaculture" edited by T.V.R. Pillay and Wm. A. Dill, FAO and fishing News Books Ltd. 1979: 127-134.

H. Chaudhri & J.V. Juario 1977, Use of hormones in breeding cultivated warmwater fishes with special reference to milkfish *Chanos chanos* (Forskal), *Fish. Journ. Philippines*, 11(2) 1977: 1-6.

H. Chaudhari & S. B. Singh 1983, Tilapia Culture. Dept. of Livestock: FAO/UNPD Project, Vientiane, Nov. 1983: 9p.

H. Chaudhari & S. B. Singh 1984, Induced Breeding of Carps. ICAR, New Delhi. 1984 83p.

H. Chaudhari & S.B. Singh 1984, Carp Culture. Department of Livestock; FAO/UNDP Project, Vientiane, Sept 1984: 13p.

H. Chaudhari & S.B. Singh 1985, Induced Breeding of Asiatic Carps. Depatment of Livestock FAO/UNDP Project, Vientiane, April 1985: 27p.

H. Chaudhari & S.B. Singh 1985, Handbook on Fish Culture. Depatment of Livestock. FAO/UNDP Project, Vientiane, April 1985: 27p.

H. Chaudhari *et al.* 1962, Observations on induced breeding of major Indian carps during 1960 at Cuttack. Proc. Indian Sci Congr. 49 (3): 391.

H. Chaudhari *et al.* 1966, Experiments on large-scale production of seed of the Chinese grass carp. *Ctenopharyngodon idellus* (C. & V.) and the silver carp, *Hypophthalmichthys militrix* (C.&V.) by induced breeding in ponds in India, Proc. Indian Acad, Sci. Sec. B 63: 80-95.

H. Chaudhari *et al.* 1967, Note on natural spawning of grass carp and Silver carp in induced breeding experiments. *Sci & Cult.*, 33 (Nov.) 493-494.

H. Chaudhari *et al.* 1973, Role of Chinese grass carp. *Ctenopharyngodon idella* (Vol.) in biological control of noxious aquatic weeds in India: a review. Proc. Regional Seminar on noxious aquatic vegetaion, New Delhi, 12-17 Sept. 1973. In "Aquatic weeds in southeast Asia" edited by C.K. Varshney & J. Rzosks. published by W. Junk B.V., The Hague, 1976: 396p.

H. Chaudhari *et al.* 1974, Record fish production with intensive culture of Indian and exotic carps. Curr. Sci. 43 (10) May 1974 : 302-304.

H. Chaudhari *et al.* 1975, A new high in fish production in India with record yields by composite fish culture in freshwater ponds. Aquaculture 6 (1975): 343- 355.

H. Chaudhari *et al.* 1976, Notes on external sex characters of *Chanos chanos* (Forskal). *Fish. Res. Journ. Philippines,* 1 (2) 1976: 76-80.

H. Chaudhari *et al.* 1977, Artificial fertilization of eggs and early development of the milkfish, *Chanos chanos* (Forskal). Technical Report No. 3., Contribution No. 12 SEAFDEC AQO, Tigbauan, Iloilo, Philippines, Aug. 1977: 21-38.

H. Chaudhari *et al.* 1977, Experiments on artifical propagation and larval development of *Mugil cephalus* Linnaeus in India. *J. Inland fish. Soc. India.* 9 (Dec.) 1977 : 30-41.

H. Chaudhari *et al.* 1977, Efficacy of partially purified salmon gonadotrophin (S.G - G100) in inducing spawning in Indian major carps. *J. Inland Fish Soc. India,* 9 (Dec.) 1977: 163-164.

H. Chaudhari *et al.* 1977, Studies on the effect of grass carp on paddy plantation. *Sci & cult., 43* (July) 1977: 321-323

H. Chaudhari *et al.* 1978, Observations on artifical fertilization of eggs and the embryonc and larval development of milkfish, *Chanos chanos* (Forskal). *Aquaculture. 13* (1978) : 95-113.

H. Chaudhari *et al.* 1978, Record fish production through intensive fish culture in a farmer's ponds. *J. Inland Fish. Soc. India. 10* (Dec.) 1978: 19-27.

Alikunhi K. H. & H. Chaudhri 1951, On the occurrence of *Labeo rohita* in the Godavari river system. Sci & Culture., 16 (May) 1951 : 527.

H. Chaudhuri & V. Ramachandran 1951 A possibel method for segregation of species from a mixed collection of carp fry. *Sci. & Cult. 17* (5) : 220.

V. Ramachandran & H. Chaudhuri 1951, Mortality of carp fry under super saturation of dissolved oxygen in water. *Proc. Nat. Inst. Sci.* India. 17(4) : 261-264.

V. Ramachandran & H. Chaudhuri 1952, On the role of the duckweed *Lemna minor,* in the preparation of carp nurseries. *Sci. & Cult.,* 17 (April) 1952: 1-4

H. Chaudhuri & V. Ramachandran 1952, Response to transplantation of fishes in India with special reference to condition of existence of carp fry. *J. Asiat, Soc.* (sci.) 18 (1) : 35-53.

H. Chaudhuri 1953, Characteristic colouration of juvenile specimens of *Labeo (Morulius) angara* with notes on growth. *J. Bomb. Nat. Hist. Soc.,* 51 (3) : 537-542.

H. Chaudhuri & V. Ramachandran 1954, On the role of Plankton in the survival and growth of carp fry in nursery ponds. *Symp. Marine & Freshwater Plankton Indo-Pacific.* 1954 : 1-4.

H. Chaudhuri 1954, On the life-history and bionomics of the carp minnow, *Chela phulo* (Hamilton). *Proc. Indian Acad. Sci.,* 39 (2) Sec.B. 1954 : 76-90.

H. Chaudhuri & V. Ramachandran 1955, On the mortality of carp fry in nursery ponds and the role of Plankton in their Survival and growth. *Indian J. Fish.* 2 (20 195 : 257-313.

H. Chaudhuri & K. H. Alikunhi 1959, Preliminary observations on hybridization of the common carp (*Cyprinus carpio*) with Indian carps. *Proc. Indian Sci Congr.*, 46 (4).

Ibrahim K. H. & H. Chaudhuri 1960, Preservation of fish pituitary extract in glycerine in ampoules for commercial production. *Indian J. Exp. Biol.4* (4) 249-250.

H. Chaudhuri 1970, Studies on the role of sex-specificity of carp pituitary glands in induced spawning in Indian carps. *J. Inland Fish. Soc. India* 2 (1970).

Bhowmick, R. M & H. Chaudhuri 1968, Preliminary observations on the use of Trichloracetic acid for extraction of gonadotropic hormones in induced spawning of carps. *J. Zool. Soc. India.* 20 (1&2): 48-54.

H. Chaudhuri 1975, Report on the progress of freshwater Sciences in India during 1970-74. Formal Proceedings 15th Pact. Sci. Congr. 1975, Vancouver B. C. 2: 107-112.

Kreuze J.F.P., H. Chaudhuri, L. Hartmann & P. Haberlin 1977, Feasibility study on Fish Farm at Nabong in the Lao PDR. Report of the Committee for Coordination of Investigations of the Lower Mekong Basin, BKK, Feb. 1977: 113p. (mimeo).

Bhimachar B.S., G.G. Banerjee, P. Mahapatra, H. Chaudhuri, S. Sharma, J.V.H. Dixitulu, G.P. Dubey & H.P.C. Shetty 1966, Report of the Fish Seed Committee, Govt. of India, Ministry of Food and Agriculture (Department of Agriculture), 1966: 212 p.

H. Chaudhuri 1972, Role of predatory aquatic insets in carp nurseries, together with notes on the life-history stages of four common aquatic beetles. Paper presented on the Silver Jubilee Symposia on "Aquaculture as an Industry: CIFRI (ICAR): Barrackpore, Nov. 25-28, 1972 (mimeo). Abstract published. 1972: 19-20.

H. Chaudhuri 1994, History of induced breeding in fishes and its application to Aquaculture. *Proc. Zool. Soc.*, 47(1): 1-31. 1994.

H. Chaudhuri 1994, *et al.* Ipomoea carnea Jacq. A new aquatic weed problem in India. (Regional Seminar on 'Noxious Aquatic vegetation. New Delhi. 12-17 Sept. 1973. *J. Aquat. Plant Management*. 32. 37-38.

INDEX